FUNCTIONAL
CALCULI

FUNCTIONAL CALCULI

Carlos Bosch
Instituto Tecnológico Autónomo de México, México

Charles Swartz
New Mexico State University, USA

 **World Scientific**

NEW JERSEY · LONDON · SINGAPORE · BEIJING · SHANGHAI · HONG KONG · TAIPEI · CHENNAI

Published by

World Scientific Publishing Co. Pte. Ltd.
5 Toh Tuck Link, Singapore 596224
USA office: 27 Warren Street, Suite 401-402, Hackensack, NJ 07601
UK office: 57 Shelton Street, Covent Garden, London WC2H 9HE

British Library Cataloguing-in-Publication Data
A catalogue record for this book is available from the British Library.

FUNCTIONAL CALCULI

ISBN 978-981-4415-97-2

Printed in Singapore by World Scientific Printers.

Preface

This book contains an exposition of several functional calculi and the background material necessary for these developments. Roughly speaking, a functional calculus is a construction which associates with an operator or a family of operators a homomorphism from a function space into a subspace of continuous linear operators. That is, a method for defining "functions of an operator". The simplest example of such a functional calculus is the case when $\mathbf{A}$ is a continuous linear operator on a Banach space X and p, $p(z) = \sum_{k=0}^{n} a_k z^k$, is a polynomial in one variable. Then $p(\mathbf{A})$ has a natural definition as

$$p(\mathbf{A}) = \sum_{k=0}^{n} a_k \mathbf{A}^k$$

and this gives a homomorphism, $p \longrightarrow p(\mathbf{A})$, from the algebra $\mathscr{P}$ of polynomials into the algebra $L(X)$ of continuous linear operators on X. Any proposed functional calculus should give an extension of this homomorphism.

Perhaps the most familiar example of such a functional calculus is based on the spectral theorem for continuous linear, self adjoint operators on a Hilbert space. If $\mathbf{A}$ is a continuous, self adjoint operator on a complex Hilbert space H, then there is a projection valued measure E on the Borel sets of the spectrum, $\sigma(\mathbf{A})$, of $\mathbf{A}$ such that

$$\mathbf{A} = \int_{\sigma(\mathbf{A})} z \, dE(z).$$

If f is a continuous function defined on $\sigma(\mathbf{A})$, we may define a functional calculus by setting

$$f(\mathbf{A}) = \int_{\sigma(\mathbf{A})} f(z)\, dE(z).$$

This functional calculus defined on $\mathcal{C}\big(\sigma(\mathbf{A})\big)$, the space of continuous functions on $\sigma(\mathbf{A})$, extends the functional calculus $p \longrightarrow p(\mathbf{A})$ for polynomials and is described in Chapter 2.

In the first chapter we present the background on vector and operator valued measures which is necessary for the development of several of the functional calculi. The Orlicz–Pettis Theorem is an important tool in this development and is covered in an appendix.

In the second chapter we present the most familiar of the functional calculi based on the spectral theorem for continuous, self adjoint operators on a Hilbert space which is described above.

In Chapter 3 this functional calculus is extended to the case of several continuous, commuting, self adjoint operators and in Chapter 4 is extended to normal operators.

In Chapter 5 we develop the integration theory for vector valued functions needed in the final two chapters. We define the Pettis and Bochner integrals and establish their basic properties.

Chapter 6 describes some of the desired properties of a functional calculus.

Chapter 7 contains an exposition of the Riesz operational calculus. The Riesz operational calculus is based on the Cauchy integral formula,

$$f(z) = \frac{1}{2\pi i} \int_C \frac{f(w)}{(w - z)}\, dw.$$

If $\mathbf{A}$ is a continuous linear operator on a complex Banach space X and f is an analytic function defined on an open set containing $\sigma(\mathbf{A})$, $f(\mathbf{A})$ is defined by formally replacing $(w - z)^{-1}$ in the Cauchy integral formula by $(w - \mathbf{A})^{-1}$. We show that this formal definition is meaningful and extends the functional calculus for polynomials. The basic properties of the calculus are then developed.

Chapter 8 gives an exposition of a functional calculus due to H. Weyl which has its origins in quantum mechanics. The calculus is first defined for a continuous linear operator and then extended to the case of several continuous linear operators which do not necessarily commute.

The background needed to follow the exposition include a knowledge of the Lebesgue integral and basic properties of continuous linear operators on Banach spaces. In Chapters 7 and 8 some acquaintance with Fréchet and locally convex spaces is needed. There are appendices which contain some of the more specialized topics which are required. These contain the Orlicz–Pettis Theorem needed for measure and integration, a description of the spectral properties of operators and an appendix on Fourier transforms of tempered distributions including the Paley–Wiener theorem which is needed in Chapter 8.

We want to thank the nice environment created by the Department of Mathematical Sciences at New Mexico State University where we wrote the first manuscript. The first author acknowledges the support of the Asociación Mexicana de Cultura.

Carlos Bosch
Charles Swartz

Contents

Chapter 1

Vector and Operator Valued Measures

In this chapter we record some of the measure and integration results for vector valued measures needed in subsequent chapters. For an introduction to measure theory, see [Roy] or [Sw1].

1.1 Vector Measures

Let $\mathcal{A}$ be an algebra of subsets of a set S and let Σ be the σ-algebra generated by $\mathcal{A}$. Let X be a Banach space. Let us recall that a set function $\mu : \mathcal{A} \to X$ is *finitely additive* if

$$\mu(A \bigcup B) = \mu(A) + \mu(B)$$

for $A, B \in \mathcal{A}$, $A \cap B = \emptyset$. The set function μ is *countably additive* if

$$\mu(\bigcup_{j=1}^{\infty} A_j) = \sum_{j=1}^{\infty} \mu(A_j)$$

when $\{ A_j \} \subset \mathcal{A}$ is pairwise disjoint and $\bigcup_{j=1}^{\infty} A_j \in \mathcal{A}$. Note that the series $\sum_{j=1}^{\infty} \mu(A_j)$ is unconditionally convergent in the sense that any rearrangement of the series still converges to $\sum_{j=1}^{\infty} \mu(A_j)$ since the union does not depend on the arrangement, or ordering, of the terms.

A countably additive set function $\mu : \Sigma \to X$ is called a *vector measure*. There is a very convenient test for countable additivity for finitely additive set functions $\mu : \Sigma \to X$ defined on σ-algebras, called the Orlicz–Pettis Theorem; namely, such a function μ is countably additive if and only if the scalar set function $x'\mu = x' \circ \mu$ is countably additive for every $x' \in X'$. See Appendix A for details.

Proposition 1.1. *If $\mu : \Sigma \to X$ is countably additive, then μ has bounded range.*

1

Proof. For $x' \in X'$, $x'\mu$ is a countably additive scalar measure and, therefore, has bounded range ([Sw1] 2.2.1.5). The result now follows from the Uniform Boundedness Principle ([Sw1] 5.3.1). □

It is actually the case that the range of a vector measure is relatively weakly compact, but we do not need this result in what follows (See [DU] I.2.7).

If $\mu : \mathcal{A} \to X$ is finitely additive, the (scalar) *semi-variation* of μ is defined to be

$$\widehat{\mu}(A) = \sup \left\{ \left\| \sum_{j=1}^{n} a_j \mu(A_j) \right\| \right\},$$

where the supremum is taken over all the finite partitions

$$\{A_j : 1 \leq j \leq n\} \subset \mathcal{A}$$

of A and all scalars $|a_j| \leq 1$.

Proposition 1.2. *Let* $\mu : \Sigma \to X$ *be countably additive. Then for* $A \in \Sigma$,

$$\sup \{\|\mu(B)\| : B \subset A, B \in \Sigma\} \leq \widehat{\mu}(A) \leq 4 \sup\{\|\mu(B)\| : B \subset A, B \in \Sigma\}.$$

Proof. Clearly $\|\mu(B)\| \leq \widehat{\mu}(A)$ when $B \subset A$ and $B \in \Sigma$, so the first inequality follows. For the second inequality we use an inequality for signed measures:

If $\alpha : \Sigma \to \mathbb{C}$ is countably additive, then its variation $|\alpha|$ is given by

$$|\alpha|(A) = \sup \left\{ \sum_{j=1}^{n} |\alpha(A_j)| : \{A_j\} \subset \Sigma, \text{ a partition of } A \right\}$$

and we have

$$\sup \{|\alpha(B)| : B \subset A, B \in \Sigma\} \leq |\alpha|(A) \leq 4 \sup\{|\alpha(B)| : B \subset A, B \in \Sigma\},$$

([DS] III.1.5, [Sw1] 2.2.1.7). Thus,

$$\widehat{\mu}(A) = \sup \left\{ \left\| \sum_{j=1}^{n} a_j \mu(A_j) \right\| \right\} = \sup_{\|x'\| \leq 1} \sup \left\{ \left| \sum_{j=1}^{n} a_j x' \mu(A_j) \right| \right\}$$

$$\leq \sup_{\|x'\| \leq 1} \sup \left\{ \sum_{j=1}^{n} |a_j| \, |x'\mu|(A_j) \right\} \leq \sup_{\|x'\| \leq 1} \{|x'\mu|(A)\}$$

$$\leq 4 \sup_{\|x'\| \leq 1} \sup_{B \subset A} \{|x'\mu(B)|\} \leq 4 \sup_{B \subset A} \{\|\mu(B)\|\},$$

for each $A \in \Sigma$. □

We next define the integral of a bounded scalar function with respect to a vector measure.

Let $\mu : \Sigma \to X$ be a vector measure and let $s : S \to \mathbb{C}$ be Σ-simple with

$$s = \sum_{j=1}^{n} a_j \chi_{A_j} \text{ for } A_j \in \Sigma.$$

We define the integral of s with respect to μ over $A \in \Sigma$ to be

$$\int_A s d\mu = \sum_{j=1}^{n} a_j \mu(A_j \cap A).$$

As in the case of scalar measures, it is readily checked that this definition does not depend on the particular representation of s. This requires only finite additivity.

Lemma 1.3. *If* $s = \sum_{j=1}^{n} a_j \chi_{A_j}$ *is* Σ-simple and $A \in \Sigma$, then

$$\left\| \int_A s d\mu \right\| \leq \|s\|_\infty \, \widehat{\mu}(A), \tag{1.1}$$

where $\|s\|_\infty = \sup\{|s(t)| : t \in S\}$.

Proof. We may assume that the family $\{A_j\}$ is disjoint with union S and that $s \neq 0$. Then

$$\left\| \int_A s d\mu \right\| = \|s\|_\infty \left\| \sum_{j=1}^{n} (a_j / \|s\|_\infty) \mu(A_j \cap A) \right\|$$

$$\leq \|s\|_\infty \, \widehat{\mu}(A). \qquad \square$$

Now suppose $f : S \to \mathbb{C}$ is bounded and Σ-measurable and put

$$\|f\|_\infty = \sup\{|f(t)| : t \in S\} \, .$$

Pick a sequence of Σ-simple functions $\{s_j\}$ which converge uniformly to f on S ([Sw1] 3.1.1.3). From (1.1) it follows that $\{\int_A s_j d\mu\}$ is a Cauchy sequence in X for every $A \in \Sigma$. Therefore, we may define $\int_A f d\mu$ to be

$$\lim \int_A s_j d\mu = \int_A f d\mu.$$

Note that this definition is independent of the sequence $\{s_j\}$ converging uniformly to f. Indeed, if $\{g_j\}$ is another sequence of simple functions converging uniformly to f, consider the interlaced sequence $\{s_1, g_1, s_2, g_2, \ldots\}$.

Note that (1.1) also holds for f,

$$\left\| \int_A f d\mu \right\| \le \|f\|_\infty \, \widehat{\mu}(A).$$

We have the following relation to scalar integrals.

Proposition 1.4. *Let $f : S \to \mathbb{C}$ be bounded, Σ-measurable and let $x' \in X'$, $A \in \Sigma$. Then*

$$\left\langle x', \int_A f d\mu \right\rangle = \int_A f d\left(x'\mu\right).$$

Proof. If f is a simple function, the equality follows directly from the definition of the integral. If $\{s_j\}$ is a sequence of simple functions converging uniformly to f, then

$$\left\langle x', \int_A f d\mu \right\rangle = \lim \left\langle x', \int_A s_j d\mu \right\rangle = \lim \int_A s_j d\left(x'\mu\right) = \int_A f d\left(x'\mu\right),$$

since $x'\mu$ is a bounded scalar measure from Proposition 1.1. $\qquad\square$

Corollary 1.5. *Let $f : S \to \mathbb{C}$ be bounded and Σ-measurable. Then the indefinite integral*

$$A \to \int_A f d\mu$$

from Σ into X is a vector measure.

Proof. This follows from Proposition 1.4 and the Orlicz–Pettis Theorem (A.6) as described above. $\qquad\square$

1.2 Operator Valued Measures

Let $L(X)$ be the space of all continuous linear operators from X into itself. We consider integrating with respect to finitely additive set functions $E : \Sigma \to L(X)$ which are not vector measures in the sense described above but are only countably additive with respect to the strong operator topology. That is, we assume that the set function $E_x : \Sigma \to X$ defined by $E_x(A) = E(A)x$ is a vector measure, that is, countably additive, for every $x \in X$. We shall refer to such set functions as *operator valued measures*.

Proposition 1.6. *If $E : \Sigma \to L(X)$ is an operator valued measure, then E has bounded range.*

Proof. For each $x \in X$, $\{E_x(A) : A \in \Sigma\}$ has bounded range by Proposition 1.1, so the result follows from the Uniform Boundedness Principle ([Sw1] 5.3.1). $\qquad\square$

We now define the integral of a bounded, measurable function with respect to an operator valued measure $E : \Sigma \to L(X)$. Let $f : S \to \mathbb{C}$ be bounded and Σ-measurable. If $x \in X$, then $\int_A f dE_x$ exists for every $A \in \Sigma$ and by (1.1) and Proposition 1.2,

$$\left\| \int_A f dE_x \right\| \leq \|f\|_\infty \widehat{E_x}(A) \leq 4 \|f\|_\infty \sup \{\|E_x(B)\| : B \subset A, B \in \Sigma\}$$

$$\leq 4 \|f\|_\infty \|x\| \sup\{\|E(B)\| : B \subset A, B \in \Sigma\}.$$

Thus, $x \to \int_A f dE_x$ defines a continuous linear operator from X into itself. Denote this operator by $\int_A f dE$, so we have

$$\left(\int_A f dE \right) x = \int_A f dE_x$$

and

$$\left\| \int_A f dE \right\| \leq 4 \|f\|_\infty \sup \{\|E(B)\| : B \subset A, B \in \Sigma\}.$$

Notice that there is an ambiguity in notation here, since we have used $\int_A f d\mu$ and $\int_A f dE$ for integrals with respect to different types of measures μ and E. However, it should be clear from the context which type of integral we are employing.

Proposition 1.7. *For $x \in X$, $x' \in X'$ let $E_{x'x}$ be the scalar measure defined by $E_{x'x}(A) = \langle x', E(A)x \rangle = \langle x', E_x(A) \rangle$. Then*

$$\left\langle x', (\int_A f dE)x \right\rangle = \int_A f dE_{x'x}.$$

Proof. This follows from Proposition 1.4. $\qquad\square$

Proposition 1.8. *Let $f : S \to \mathbb{C}$ be bounded and Σ-measurable. Then the indefinite integral, $\int f dE$, from Σ into $L(X)$ defined by*

$$(\int f dE)(A) = \int_A f dE,$$

is an operator valued measure. Moreover, if $T \in L(X)$ commutes with every operator $E(A)$ for $A \in \Sigma$, then T commutes with $\int_A f dE$.

Proof. Corollary 1.5 gives the first statement. The last statement is readily checked for simple functions and then the general case follows by taking limits. $\qquad\square$

For later use, we record here some properties of the integral when the operator valued measure has special properties.

Let Ω be a σ-compact, locally compact Hausdorff space. That is, Ω is a countable union of compact subsets. Let $\mathcal{B}(\Omega)$ indicate the σ-algebra of Baire sets of Ω. Since we assume that Ω is σ-compact, $\mathcal{B}(\Omega)$ is the σ-algebra generated by the compact $\mathcal{G}_\delta$ sets in Ω ([Ha], [Be1]). Finally, let H be a complex Hilbert space.

Definition 1.9. A set function E defined on $\mathcal{B}(\Omega)$ is a spectral measure if

(a) E is an operator valued measure with values in $L(H)$,
(b) $E(A)$ is an orthogonal projection for every $A \in \mathcal{B}(\Omega)$,
(c) $E(\emptyset) = 0$, $E(\Omega) = I$ and
(d) $E(A \cap B) = E(A)E(B) = E(B)E(A)$ for $A, B \in \mathcal{B}(\Omega)$.

Remark 1.10. In order to avoid technical difficulties later, we choose the class of Baire sets instead of the Borel sets. In all of the applications, Ω will be a subset of $\mathbb{C}$ so the families of Baire sets and Borel sets coincide in this case ([Ha], [Be1]). Since every Baire measure is regular ([Ha], [Be1]), for every $x, y \in H$ and $A \in \mathcal{B}(\Omega)$ the signed measure

$$E_{yx}(A) = y \cdot E(A)x$$

is regular.

Let us recall that a function $f : \Omega \to \mathbb{C}$ is called a Baire function if it is measurable with respect to the Baire σ-algebra $\mathcal{B}(\Omega)$.

Example 1.11. Let $H = L^2(\mathbb{R})$ with respect to the Lebesgue measure. For $A \in \mathcal{B}(\mathbb{R})$ define $E(A) : H \to H$ by

$$E(A)f = \chi_A f.$$

It is readily checked that E is a spectral measure.

Example 1.12. Again, let $H = L^2(\mathbb{R})$ with respect to the Lebesgue measure. For $r \in \mathbb{R}$ define $E : \mathcal{B}(\mathbb{R}) \to L(H)$ by $E(A) = I$ if $r \in A$ and $E(A) = 0$ if $r \notin A$. Then E is a spectral measure.

Theorem 1.13. *Let $E : \mathcal{B}(\Omega) \to L(H)$ be a spectral measure and let $f, g : \Omega \to \mathbb{C}$ be bounded Baire functions. Then*

(i) $\left\| \int_A f dE \right\| \leq \|f\|_\infty$ *for every $A \in \mathcal{B}(\Omega)$,*
(ii) $\left(\int_A f dE \right) \left(\int_A g dE \right) = \int_A fg dE$ *for $A \in \mathcal{B}(\Omega)$. That is, the product of the integrals is the integral of the product!*

(iii) $(\int_A f dE)^* = \int_A \bar{f} dE,$

(iv) *for every* $x \in H$, $\int_A |f|^2 dE_{xx} = \left\| (\int_A f dE)x \right\|^2,$

(v) *if* $\{f_j\}$ *is a uniformly bounded sequence of Baire functions which converge pointwise to* f, *then* $\{\int_A f_j dE\}$ *converges to* $\int_A f dE$ *in the strong operator topology*,

(vi) *if* $T \in L(H)$ *commutes with each* $E(A)$ *for* $A \in \mathcal{B}(\Omega)$, *then* T *commutes with* $\int_A f dE$.

Proof. (i): First suppose that f is a simple function with $f = \sum_{j=1}^n a_j \chi_{A_j}$ and the sets $\{A_j\}$ disjoint. Using properties (a) through (d) of Definition 1.9, if $x \in H$, we have

$$\left\| \left(\int_A f dE \right)x \right\|^2 = \sum_{i=1}^n a_i E(A \cap A_i)x \cdot \sum_{j=1}^n a_j E(A \cap A_j)x$$

$$= \sum_{j=1}^n |a_j|^2 \, E(A \cap A_j)x \cdot x \leq \|f\|_\infty^2 \sum_{j=1}^n E(A \cap A_j)x \cdot x$$

$$= \|f\|_\infty^2 \, E(A)x \cdot x \leq \|f\|_\infty^2 \|x\|^2 \,.$$

So, the result holds for simple functions. Then the result holds also for bounded functions by the definition of the integral.

(ii): For convenience, set

$$S = \int_A f dE, \quad T = \int_A g dE.$$

For $x, y \in H$, define a scalar measure μ on $\mathcal{B}(\Omega)$ by setting

$$\mu(A) = E(A)Tx \cdot y.$$

From (b) and (d) in Definition 1.9, we have

$$\mu(A) = Tx \cdot E(A)y = \left(\int_A g dE \right) x \cdot E(A)y$$

$$= \left(\int_A g(t) dE(t) \right) x \cdot E(A)y)$$

$$= \int_A g(t) d\left(E(t)x \cdot E(A)y \right) = \int_A g(t) d\left(E(A) E(t) x \cdot y \right)$$

$$= \int_A g(t) d\left(E\left(t \cap A \right) x \cdot y \right),$$

where $E(t \cap A)$ denotes the restriction the spectral measure E to the restricted σ-algebra $\mathcal{B}(\Omega) \cap A$. Hence,

$$STx \cdot y = \int_A f(t) d(E(t)Tx \cdot y) = \int_A f(t) d\mu(t) = \int_A f(t)g(t) d(E(t)x \cdot y),$$

so,

$$\left(\int_A f dE\right)\left(\int_A g dE\right) = \int_A fg dE.$$

(iii): If $f = \sum_{j=1}^n a_j \chi_{A_j}$ is a simple function as in the proof of (i),

$$\left(\int_A f dE\right)^* = \left(\sum_{j=1}^n a_j E(A \bigcap A_j)\right)^* = \sum_{j=1}^n \overline{a_j} E(A \bigcap A_j) = \int_A \overline{f} dE.$$

The result for bounded f follows from the definition of the integral.

(iv): From (ii) and (iii),

$$\left\|\left(\int_A f dE\right)x\right\|^2 = \left(\int_A f dE\right)^*\left(\int_A f dE\right)x \cdot x = \left(\int_A f\overline{f} dE\right)x \cdot x = \int_A |f|^2 dE_{xx}.$$

Finally, (v) follows from (iv) and the Dominated Convergence Theorem, while (vi) follows from Proposition 1.8. $\qquad\square$

Remark 1.14. From (ii) and (iii) in Theorem 1.13, $\int_A f dE$ is a normal operator and it is self adjoint if f is real valued.

Let $B(\Omega)$ be the space of all bounded Baire functions $f : \Omega \to \mathbb{C}$. Then $B(\Omega)$ is a Banach algebra under the usual operations of pointwise sums and products and the sup-norm,

$$\|f\|_\infty = \sup\{|f(t)| : t \in \Omega\}.$$

If $E : B(\Omega) \to L(H)$ is a spectral measure, we then have a natural map $\Psi : B(\Omega) \to L(H)$ defined by

$$\Psi f = \int_\Omega f dE.$$

From Theorem 1.13 we have:

Theorem 1.15. *The map* $\Psi : B(\Omega) \to L(H)$ *is an algebra isomorphism which is continuous with*

$$\|\Psi f\| \le \|f\|_\infty$$

for $f \in B(\Omega)$. *Moreover,*

$$(\Psi f)^* = \int_\Omega \overline{f} dE$$

for $f \in B(\Omega)$.

We also have a "change of measure" principle which will be used later.

Theorem 1.16. *Let $E : \mathcal{B}(\Omega) \to L(H)$ be a spectral measure. Let Λ be a σ-compact, locally compact Hausdorff space and let $h : \Omega \to \Lambda$ be a homeomorphism. Let $F : \mathcal{B}(\Lambda) \to L(H)$ be the spectral measure defined by*

$$F(A) = E(h^{-1}(A)).$$

If $f : \Lambda \to \mathbb{C}$ is a bounded Baire function, then

$$\int_\Lambda f \, dF = \int_\Omega f \circ h \, dE.$$

Proof. It is routine to check that F is a spectral measure and that the desired change of variable formula holds when f is a simple Baire function. The general formula then follows from the definition of the integral. □

Proposition 1.17. *If $E : \mathcal{B}(\Omega) \to L(H)$ is a spectral measure, the following statements hold:*

(i) E is monotone, i.e., $A, B \in \mathcal{B}(\Omega)$ and $A \subset B$ imply $E(A) \le E(B)$,

(ii) E is subadditive, i.e., if $\{A_j\}_{j=1}^n \subset \mathcal{B}(\Omega)$, then

$$E(\bigcup_{j=1}^n A_j) \le \sum_{j=1}^n E(A_j).$$

Proof. Both, (i) and (ii), follow easily from the additivity of E and the fact that $E(A) \ge 0$ for $A \in \mathcal{B}(\Omega)$. □

Let us mention that since the values of a spectral measure are positive operators, we can define the support of such a measure and sometimes reduce the region of integration when integrating with respect to a spectral measure. In fact, let

$$C(E) = \bigcup \{U : U \text{ is open and } E(U) = 0\}.$$

The support of E, $supp(E)$, is defined to be the complement of $C(E)$ in Ω.

Lemma 1.18. *If K is compact and $K \subset C(E)$, then $E(K) = 0$.*

Proof. For each $t \in K$ there is an open set $U_t \subset C(E)$ such that $t \in U_t$ and $E(U_t) = 0$. Therefore, there exist open sets $U_1, ..., U_n \subset C(E)$ such that $K \subset \bigcup_{j=1}^n U_j$. Then $E(K) = 0$ by Proposition 1.17. □

Corollary 1.19. *If $E : \mathcal{B}(\Omega) \to L(H)$ is a spectral measure, the following statements hold:*

(i) $E(C(E)) = 0$,

(ii) $E(A) = E(A \cap supp(E))$ *for* $A \in \mathcal{B}(\Omega)$,

(iii) *if* $f : \Omega \to \mathbb{C}$ *is a bounded Baire function, then*

$$\int_A f dE = \int_{A \cap supp(E)} f dE$$

for $A \in \mathcal{B}(\Omega)$.

Proof. (i): For $x, y \in H$, $E_{yx}(C(E)) = 0$ by Lemma 1.18 and the regularity of the signed measure E_{yx}. Hence, (i) holds.

(ii) follows from (i), while (iii) follows from (i) when f is a simple Baire function. The general result is then immediate. $\square$

1.3   Extensions of Measures

We present an extension result for operator valued measures which will be used in later chapters.

Let $\mathcal{A}$ be an algebra of subsets of a set S and let Σ be the σ-algebra generated by $\mathcal{A}$.

Lemma 1.20. *If* $\mu : \mathcal{A} \to \mathbb{C}$ *is bounded and countably additive, then* μ *has a unique countably additive extension* $\widehat{\mu} : \Sigma \to \mathbb{C}$ *such that*

$$\sup\{|\mu(A)| : A \in \mathcal{A}\} = \sup\{|\widehat{\mu}(A)| : A \in \Sigma\}.$$

Proof.   First assume $\mu : \mathcal{A} \to \mathbb{R}$ and let $\mu = \mu^+ - \mu^-$ be the Jordan decomposition of μ ([Sw1] 2.2.1.3). Then extend each μ^+ and μ^- to finite positive measures on Σ, $\widehat{\mu^+}$, $\widehat{\mu^-}$, respectively and finally, put $\widehat{\mu} = \widehat{\mu^+} - \widehat{\mu^-}$.

On the other hand, if $\mu : \mathcal{A} \to \mathbb{C}$, write $\mu = \mu_1 + i\mu_2$, where the functions μ_j are real valued. Apply the result above to μ_1 and μ_2 and set $\widehat{\mu} = \widehat{\mu}_1 + i\widehat{\mu}_2$. Put

$$a = \sup\{|\mu(A)| : A \in \mathcal{A}\}$$

and let

$$\mathcal{M} = \{A \in \Sigma : |\widehat{\mu}(A)| \leq a\}.$$

Then $\mathcal{A} \subset \mathcal{M}$ and it is readily checked that $\mathcal{M}$ is a monotone class since $\widehat{\mu}$ is countably additive. Therefore, by the Monotone Class Lemma ([Sw1] 2.1.15), $\mathcal{M} = \Sigma$.

Uniqueness follows from standard scalar results ([Sw1] 2.4.12). $\square$

We now use Lemma 1.20 to obtain an extension theorem for Hilbert space valued set functions.

Theorem 1.21. *Let H be a Hilbert space and let $\mu : \mathcal{A} \to H$ be bounded and such that the set function*

$$\mu_y : \mathcal{A} \to \mathbb{C}, \text{ with } \mu_y(A) = y \cdot \mu(A)$$

is countably additive for every $y \in H$. Then μ has a unique, bounded, countably additive extension $\widehat{\mu} : \Sigma \to H$ such that

$$\sup\{\|\mu(A)\| : A \in \mathcal{A}\} = \sup\{\|\widehat{\mu}(A)\| : A \in \Sigma\}.$$

Proof. By Lemma 1.20 each μ_y has a unique, countably additive extension $\widehat{\mu_y} : \Sigma \to \mathbb{C}$. Let

$$\mathcal{M} = \{A \in \Sigma : \text{there exists } y_A \in H \text{ such that } \widehat{\mu_y}(A) = y \cdot y_A \text{ for all } y \in H\}.$$

Clearly, $\mathcal{A} \subset \mathcal{M}$ and if $A, B \in \mathcal{M}$, $A \subset B$, then $B \backslash A \in \mathcal{M}$ with $y_{B\backslash A} = y_B - y_A$.

We claim that $\mathcal{M}$ is a monotone class. For this suppose $\{A_j\} \subset \mathcal{M}$ with $A_j \uparrow A$. Set $A_0 = \emptyset$ and $B_j = A_j \backslash A_{j-1}$ for $j \geq 1$. Then

$$\widehat{\mu_y}(A_j) = y \cdot y_{A_j} = y \cdot \sum_{i=1}^{j} y_{B_i} \to \widehat{\mu_y}(A)$$

by countable additivity. Therefore, the partial sums of the series $\sum_{j=1}^{\infty} y_{B_j}$ are weakly Cauchy in H, and the series converges weakly to some $z \in H$. Since $\widehat{\mu_y}(A) = y \cdot z$, we have $z = y_A$ so $A \in \mathcal{M}$.

If $\{B_j\} \subset \mathcal{M}$ and $B_j \downarrow B$, then $S \backslash B_j \uparrow S \backslash B$ so $S \backslash B \in \mathcal{M}$ and $B \in \mathcal{M}$ by the observation above. Hence, $\mathcal{M}$ is a monotone class and by the Monotone Class Lemma ([Sw1] 2.1.15), $\mathcal{M} = \Sigma$. Then, $\widehat{\mu} : \Sigma \to H$ defined by $\widehat{\mu}(A) = y_A$ defines a finitely additive set function such that $y \cdot \widehat{\mu}(A) = \widehat{\mu_y}(A)$ for $A \in \Sigma$. Then $\widehat{\mu}$ is norm countably additive by the Orlicz–Pettis Theorem (A.6).

The equality

$$\sup\{\|\mu(A)\| : A \in \mathcal{A}\} = \sup\{\|\widehat{\mu}(A)\| : A \in \Sigma\}$$

is proven as in Lemma 1.20. Uniqueness follows from the uniqueness of each $\widehat{\mu_y}$. $\square$

Finally, we present an extension theorem for operator valued set functions.

If $E : \mathcal{A} \to L(H)$ and $x, y \in H$, we define

$$E_x : \mathcal{A} \to H \text{ and } E_{yx} : \mathcal{A} \to \mathbb{C}$$

by

$$E_x(A) = E(A)x \text{ and } E_{yx}(A) = y \cdot E(A)x.$$

Theorem 1.22. *Let H be a Hilbert space and let $E : \mathcal{A} \to L(H)$ be such that E_{yx} is bounded and countably additive for every $x, y \in H$. Then, there exists a unique operator valued measure $\widehat{E} : \Sigma \to L(H)$ extending E with*

$$\sup\{\|E(A)\| : A \in \mathcal{A}\} = \sup\{\left\|\widehat{E}(A)\right\| : A \in \Sigma\}.$$

Proof.   Since $\{E_{yx}(A) : A \in \mathcal{A}\}$ is bounded for every $x, y \in H$,

$$\sup\{\|E(A)\| : A \in \mathcal{A}\} = k < \infty$$

by the Uniform Boundedness Principle. Then

$$|E_{yx}(A)| \le k \|x\| \|y\|$$

for $A \in \mathcal{A}, x, y \in H$.

Now $E_x : \mathcal{A} \to H$ satisfies the hypothesis of Theorem 1.21 and, therefore, it has a unique countably additive extension $\widehat{E_x} : \Sigma \to H$ satisfying

$$\left\|\widehat{E_x}(A)\right\| \le k \|x\|$$

for $A \in \Sigma$.

For $A \in \Sigma$, we define $E(A)\hat{} : H \to H$ by $E(A)\hat{}x = \widehat{E_x}(A)$. By uniqueness, $E(A)\hat{}$ is linear and since

$$\|E(A)\hat{}x\| = \left\|\widehat{E_x}(A)\right\| \le k \|x\|,$$

$E(A)\hat{} \in L(H)$ with $\|E(A)\hat{}\| \le k$. Define $\widehat{E} : \Sigma \to L(H)$ by $\widehat{E}(A) = E(A)\hat{}$. Since $\widehat{E}(A)x = \widehat{E_x}(A)$ for $A \in \Sigma$, it follows that E is an operator valued measure satisfying $\left\|\widehat{E}(A)\right\| \le k$ for $A \in \Sigma$.

Uniqueness follows from Theorem 1.21.   □

1.4   Regularity and Countable Additivity

We prove a variant of Alexanderoff's theorem ([Sw1] 2.7.9) on regularity and countable additivity. This result will be used later in this chapter.

If Ω is a σ-compact, locally compact Hausdorff space, $\mathcal{R}$ is a subset of $\mathcal{B}(\Omega)$ and $\mu : \mathcal{R} \to \mathbb{C}$, we say that μ is *inner regular* if for every $A \in \mathcal{R}$

and $\epsilon > 0$ there exists a compact set $K \in \mathcal{R}$, $K \subset A$, such that $|\mu(B)| < \epsilon$ when $B \subset A \backslash K$, $B \in \mathcal{R}$. Let $\mathcal{A}$ be a subalgebra of $\mathcal{B}(\Omega)$.

Theorem 1.23. *Assume that Ω is compact and $\mu : \mathcal{A} \to [0, \infty)$ is inner regular. Then μ is countably additive.*

Proof. Let $\{A_j\} \subset \mathcal{A}$ be pairwise disjoint with

$$A = \bigcup_{j=1}^{\infty} A_j \in \mathcal{A}.$$

Put

$$B_n = \bigcup_{j=n}^{\infty} A_j = A \backslash \bigcup_{j=1}^{n-1} A_j \in \mathcal{A}.$$

If $\mu(B_n) \to 0$, then $\mu(A) - \sum_{j=1}^{n-1} \mu(A_j) \to 0$ and μ is countably additive. Therefore, it suffices to show that if $A_j \in \mathcal{A}$ and $A_j \downarrow \emptyset$, then $\mu(A_j) \to 0$.

Suppose there exists $\{A_j\} \subset \mathcal{A}$, $A_j \downarrow \emptyset$ with $\mu(A_j) \geq a > 0$ for every j. For each j, there exists a compact $K_j \in \mathcal{A}$, $K_j \subset A_j$ with $\mu(A_j \backslash K_j) < a/2^j$. Then

$$\mu(A_n \backslash \bigcap_{i=1}^{n} K_i) = \mu(\bigcap_{i=1}^{n} A_i \backslash \bigcap_{i=1}^{n} K_i)$$
$$\leq \mu(\bigcup_{i=1}^{n} (A_i \backslash K_i)) \leq \sum_{i=1}^{n} \frac{a}{2^i} < a,$$

so $\bigcap_{i=1}^{n} K_i \neq \emptyset$ for every n. Since Ω is compact, $\bigcap_{i=1}^{\infty} K_i \neq \emptyset$. Hence, $\bigcap_{i=1}^{\infty} A_i \neq \emptyset$. This contradiction shows that μ is countably additive. $\square$

Corollary 1.24. *Assume that Ω is compact and $\mu : \mathcal{A} \to \mathbb{C}$ is bounded and inner regular. Then μ is countably additive.*

Proof. Assume first that $\mu : \mathcal{A} \to \mathbb{R}$ and let $\mu = \mu^+ - \mu^-$ be the Jordan decomposition of μ. Note that each μ^+ and μ^- must be inner regular ([Sw1] 2.2.1) so μ is countably additive by Theorem 1.23. The result for complex valued μ then follows by considering the real and imaginary parts of μ. $\square$

1.5 Countable Additivity on Products

In this section we prove a theorem on countable additivity of set functions defined on product spaces. This result will be used when treating products of operator valued measures.

Let Ω, Λ be compact Hausdorff spaces and set $Z = \Omega \times \Lambda$. Let

$$\mathcal{R} = \{A \times B : A \in \mathcal{B}(\Omega), B \in \mathcal{B}(\Lambda)\}$$

be the semialgebra of all rectangles with Baire sets as sides and let $\mathcal{A}$ be the algebra generated by $\mathcal{R}$. Note that $\mathcal{B}(Z)$ is the σ-algebra generated by $\mathcal{A}$. If λ is a complex valued set function defined on $\mathcal{R}$, or $\mathcal{A}$ or $\mathcal{B}(Z)$, given $A \in \mathcal{B}(\Omega)($ respectively $B \in \mathcal{B}(\Lambda))$, define

$$\lambda_A : \mathcal{B}(\Lambda) \to \mathbb{C} \text{ by } \lambda_A(B) = \lambda(A \times B)$$

(respectively

$$\lambda^B : \mathcal{B}(\Omega) \to \mathbb{C} \text{ by } \lambda^B(A) = \lambda(A \times B)).$$

Theorem 1.25. *Let $\lambda : \mathcal{R} \to [0, \infty)$ be finitely additive. Assume that λ satisfies*

(a) λ_A is inner regular on $\mathcal{B}(\Lambda)$ for every $A \in \mathcal{B}(\Omega)$,
(b) λ^B is inner regular on $\mathcal{B}(\Omega)$ for every $B \in \mathcal{B}(\Lambda)$.

Then, λ is inner regular on $\mathcal{R}$.

Proof. Let $A \times B \in \mathcal{R}$ and fix $\epsilon > 0$. There exist compact sets $K \subset A, L \subset B$ such that

$$\lambda^B(A \backslash K) = \lambda((A \backslash K) \times B) < \epsilon,$$
$$\lambda_A(B \backslash L) = \lambda(A \times (B \backslash L)) < \epsilon.$$

Then $K \times L$ is compact and

$$\lambda(A \times B \backslash K \times L) = \lambda((A \backslash K) \times B) + \lambda(A \times (B \backslash L)) < 2\epsilon,$$

so λ is inner regular. $\qquad\qquad\square$

Remark 1.26. Note that if λ_A (resp. λ^B) is countably additive, then (a) (resp. (b)), in Theorem 1.25 is satisfied.

Corollary 1.27. *Let λ be as in Theorem 1.25 and let μ be the unique, finitely additive extension of λ to $\mathcal{A}$. Then, μ is inner regular on $\mathcal{A}$ and, as a consequence, it is countably additive by Theorem 1.23.*

Proof. Let $C \in \mathcal{A}$. Then $C = \bigcup_{i=1}^{n} A_i \times B_i$, where $\{A_i \times B_i\} \subset \mathcal{R}$ are pairwise disjoint. By Theorem 1.25, for each i there exists compact sets $K_i \subset A_i, L_i \subset B_i$ such that

$$\lambda(A_i \times B_i \backslash K_i \times L_i) < \epsilon/n.$$

Then $K = \bigcup_{i=1}^{n} K_i \times L_i$ is compact and the union is pairwise disjoint and contained in C. Moreover,

$$\mu(C \setminus K) = \sum_{i=1}^{n} \lambda(A_i \times B_i \setminus K_i \times L_i) < \epsilon.$$

$\square$

Theorem 1.28. *Suppose* $\lambda : \mathcal{A} \to \mathbb{C}$ *is finitely additive and satisfies*

(a) λ *is bounded on* $\mathcal{A}$,
(b) λ_A *is countably additive for each* $A \in \mathcal{B}(\Omega)$,
(c) λ^B *is countably additive for each* $B \in \mathcal{B}(\Lambda)$.

Then, λ *is countably additive on* $\mathcal{A}$.

Proof.   Assumption (a) implies that λ has finite variation, $|\lambda|$, on $\mathcal{A}$ and that $|\lambda|$ is finitely additive ([Sw1] 2.2.1). We claim that $|\lambda|$ satisfies (a) in Theorem 1.25. In fact, let $A \in \mathcal{B}(\Omega)$ and let $\{B_k\} \subset \mathcal{B}(\Lambda)$ be pairwise disjoint with $B = \bigcup B_k$. Since $|\lambda|$ is finitely additive,

$$|\lambda| (A \times B) \geq |\lambda| (\bigcup_{k=1}^{n} A \times B_k) = \sum_{k=1}^{n} |\lambda| (A \times B_k)$$

for every n. So,

$$|\lambda| (A \times B) \geq \sum_{k=1}^{\infty} |\lambda| (A \times B_k).$$

Let $\epsilon > 0$. There exist pairwise disjoint $\{C_i : 1 \leq i \leq n\} \subset \mathcal{A}$ with $A \times B = \bigcup_{i=1}^{n} C_i$ and

$$|\lambda| (A \times B) - \epsilon < \sum_{i=1}^{n} |\lambda(C_i)|. \tag{1.2}$$

Each C_i is the union of pairwise disjoint sets $\{A_j^i \times B_j^i : j = 1, ..., n_i\} \subset \mathcal{R}$ and since

$$|\lambda(C_i)| = \left| \sum_{j=1}^{n_i} \lambda(A_j^i \times B_j^i) \right| \leq \sum_{j=1}^{n_i} |\lambda(A_j^i \times B_j^i)|,$$

we may assume that each C_i in (1.2) has the form $A_i \times D_i \in \mathcal{R}$. By (b),

$$|\lambda|\,(A \times B) - \epsilon < \sum_{i=1}^{n} |\lambda\,(A_i \times D_i)|$$

$$= \sum_{i=1}^{n} \left|\lambda(A_i \times (\bigcup_{k=1}^{\infty} D_i \bigcap B_k))\right|$$

$$= \sum_{i=1}^{n} \left|\sum_{k=1}^{\infty} \lambda(A_i \times (D_i \bigcap B_k))\right|$$

$$\leq \sum_{k=1}^{\infty} \sum_{i=1}^{n} |\lambda|\,(A_i \times (D_i \bigcap B_k))$$

$$= \sum_{k=1}^{\infty} |\lambda|\,(\bigcup_{i=1}^{n} A_i \times (D_i \bigcap B_k))$$

$$\leq \sum_{k=1}^{\infty} |\lambda|\,(A \times B_k).$$

Hence,

$$|\lambda|\,(A \times B) \leq \sum_{k=1}^{\infty} |\lambda|\,(A \times B_k)$$

and we must have

$$|\lambda|\,(A \times B) = \sum_{k=1}^{\infty} |\lambda|\,(A \times B_k).$$

Therefore, $|\lambda|_A$ is countably additive and (a) in Theorem 1.25 holds by Remark 1.26.

Similarly, $|\lambda|$ satisfies (b) in Theorem 1.25 as well. Then, by Corollary 1.27, $|\lambda|$ is countably additive on $\mathcal{A}$ and since $|\lambda(A)| \leq |\lambda|\,(A)$, it follows that λ is countably additive on $\mathcal{A}$. $\square$

Theorem 1.28 is very useful in checking the countable additivity of set functions defined on products since it allows one to check countable additivity in each variable "separately". The result is used in this way in Proposition 3.3.

Remark 1.29. The theory of integration of scalar valued functions against vector valued measures that we have developed, is sufficient for what follows in the text. For a more detailed and more general presentation, see Kluvanek and Knowles ([KK]), which also has a discussion of the extension of vector measures. Theorem 1.28 is due to Kluvanek ([Kl]).

Chapter 2

Functions of a Self Adjoint Operator

We establish in this chapter the simplest of the functional calculi. We begin by defining $f(\mathbf{A})$ for any bounded self adjoint operator $\mathbf{A}$ defined on a complex Hilbert space H and any continuous, real valued function f defined on the spectrum $\sigma = \sigma(\mathbf{A})$ of $\mathbf{A}$. As applications of the functional calculus we show the existence of the square root of a positive operator and also establish the spectral representation of the operator $\mathbf{A}$. This spectral representation is then used to extend the functional calculus to the case where f is a bounded Baire function defined on σ.

Throughout this chapter $\mathbf{A}$ will be a bounded self adjoint operator defined on a complex Hilbert space H, with spectrum $\sigma = \sigma(\mathbf{A})$.

If p is a real polynomial,

$$p(t) = \sum_{j=0}^{n} a_j t^j,$$

we define

$$p(\mathbf{A}) = \sum_{j=0}^{n} a_j \mathbf{A}^j,$$

where $\mathbf{A}^0 = I$, the identity operator.

Let $\mathcal{P}$ be the space of all real polynomials; then $\mathcal{P}$ is a real algebra under the usual operations of pointwise sum and product. Note that the map $\Psi : \mathcal{P} \to L(H)$ defined as $p \to p(\mathbf{A})$, is an algebra isomorphism from $\mathcal{P}$ into $L(H)$.

Let $C(\sigma)$ be the Banach space of all continuous real valued functions defined on σ with the sup-norm,

$$\|f\|_\infty = \max\left\{|f(t)| : t \in \sigma\right\}.$$

We seek to extend the map $\Psi : \mathcal{P} \to L(H)$ to an algebra isomorphism on $C(\sigma)$. For this we have:

Lemma 2.1. *If $p(t) = \sum_{j=0}^{n} a_j t^j$ is a real polynomial, then*
$$\|p(\mathbf{A})\| = \sup\{|p(t)| : t \in \sigma\}.$$

Proof.   This follows from Theorem B.23 and Theorem C.22.   □

It follows from Lemma 2.1 that the map $\Psi : \mathcal{P} \to L(H)$ is an isometry from the subalgebra $\mathcal{P}$ of $C(\sigma)$ into $L(H)$, which is also an algebra isomorphism.

If $\mathcal{A}$ is the closed subalgebra of $L(H)$ generated by $\mathbf{A}$ or $\Psi(\mathcal{P})$, then by the Weierstrass Approximation Theorem, Ψ has a unique continuous extension, still denoted by Ψ, from $C(\sigma)$ onto $\mathcal{A}$, which is an isometry and an algebra isomorphism when $C(\sigma)$ is equipped with the usual pointwise operations.

If $f \in C(\sigma)$, we often write $f(\mathbf{A}) = \Psi(f)$, so $\|f(\mathbf{A})\| = \|f\|_\infty$. If p is a real polynomial, $\Psi(p) = p(\mathbf{A})$ as before. Thus, we have the following functional calculus.

Theorem 2.2. *The map $\Psi : C(\sigma) \to \mathcal{A}$ satisfies the properties:*

(i) $\Psi(f + g) = \Psi(f) + \Psi(g)$, $\Psi(af) = a\Psi(f)$ *for a scalar,* $\Psi(fg) = \Psi(f)\Psi(g)$ *for $f, g \in C(\sigma)$,*

(ii) $\Psi(p) = p(\mathbf{A})$ *for every $p \in \mathcal{P}$; in particular,* $\Psi(1) = I$ *and* $\Psi(t) = \mathbf{A}$,

(iii) $\|\Psi(f)\| = \|f\|_\infty$ *for $f \in C(\sigma)$,*

(iv) *if $T \in L(H)$ commutes with $\mathbf{A}$, then T commutes with every $\Psi(f)$, for $f \in C(\sigma)$.*

Remark 2.3. The inverse of the map of $\Psi : C(\sigma) \to \mathcal{A}$ is called the Gelfand map relative to $\mathbf{A}$. If we denote this map by Φ, then $\Phi : \mathcal{A} \to C(\sigma)$ is an algebra isomorphism which is also an isometry onto $C(\sigma)$.

We can now extend to continuous functions the junior grade spectral mapping theorem presented in Theorem B.23.

Theorem 2.4. *(Spectral Mapping Theorem) If $f \in C(\sigma)$, then*
$$\sigma(f(\mathbf{A})) = f(\sigma(\mathbf{A})).$$

Proof.   Suppose $\lambda \in \sigma(\mathbf{A})$. We show that $f(\mathbf{A}) - f(\lambda)I$ is not invertible, so $f(\sigma(\mathbf{A})) \subset \sigma(f(\mathbf{A}))$. In fact, choose a sequence $\{p_j\}$ of polynomials converging uniformly to f on σ. Then
$$p_j(\mathbf{A}) - p_j(\lambda)I \to f(\mathbf{A}) - f(\lambda)I$$

in norm. According to Theorem B.23, $p_j(\lambda) \in \sigma(p_j(\mathbf{A}))$. So, $p_j(\mathbf{A}) - p_j(\lambda)I$ is not invertible in $L(H)$ and according to Theorem B.11, $f(A) - f(\lambda)I$ is not invertible.

Conversely, suppose $\lambda \notin f(\sigma(\mathbf{A}))$. Then $f - \lambda$ is non-zero on $\sigma(\mathbf{A})$, so the function

$$g = \frac{1}{f - \lambda}$$

is continuous on σ. By (i) and (ii) in Theorem 2.2,

$$(f(\mathbf{A}) - \lambda I)g(\mathbf{A}) = \Psi(1) = I.$$

So $\lambda \notin \sigma(f(\mathbf{A}))$. Hence, $\sigma(f(\mathbf{A})) \subset f(\sigma(\mathbf{A}))$. $\qquad\square$

Corollary 2.5. *Let $f \in C(\sigma)$. Then $f(\mathbf{A}) \geq 0$ if and only if $f \geq 0$ on σ.*

Proof. By Theorem C.19, $f(\mathbf{A}) \geq 0$ if and only if $\sigma(f(\mathbf{A})) \subset [0, \infty)$, so the result follows from Theorem 2.4. $\qquad\square$

As another application of the functional calculus, we show that every positive operator has a unique square root.

Theorem 2.6. *Let $\mathbf{A} \in L(H)$ be positive. Then there exists a unique positive operator, denoted by $\sqrt{\mathbf{A}}$, such that $(\sqrt{\mathbf{A}})^2 = \mathbf{A}$. Moreover, if $\mathbf{A}$ is invertible, then $\sqrt{\mathbf{A}}$ is invertible.*

Proof. Since $\mathbf{A} \geq 0$, $\sigma(\mathbf{A}) \subset [0, \infty)$ and the function $f : t \to \sqrt{t}$ is continuous on $\sigma(\mathbf{A})$. Put $\sqrt{\mathbf{A}} = f(\mathbf{A})$. By Theorem 2.2,

$$(\sqrt{\mathbf{A}})^2 = f(\mathbf{A})f(\mathbf{A}) = f^2(\mathbf{A}) = \mathbf{A}.$$

For uniqueness, suppose $\mathbf{C} \geq 0$ and $\mathbf{C}^2 = \mathbf{A}$. Set $\mathbf{B} = \sqrt{\mathbf{A}}$. Then

$$\left\|\sqrt{\mathbf{C}}x\right\|^2 + \left\|\sqrt{\mathbf{B}}x\right\|^2 = \mathbf{C}x \cdot x + \mathbf{B}x \cdot x$$
$$= (\mathbf{C} + \mathbf{B})x \cdot x \quad \text{for} \quad x \in H. \qquad (2.1)$$

Since

$$\mathbf{A}\mathbf{C} = \mathbf{C}^2\mathbf{C} = \mathbf{C}\mathbf{C}^2 = \mathbf{C}\mathbf{A},$$

the operator $\mathbf{C}$ commutes with $\mathbf{B}$ by Theorem 2.2. For $y \in H$ put $x = (\mathbf{B} - \mathbf{C})y$ in (2.1). Then,

$$\left\|\sqrt{\mathbf{C}}x\right\|^2 + \left\|\sqrt{\mathbf{B}}x\right\|^2 = (\mathbf{C} + \mathbf{B})x \cdot x$$
$$= (\mathbf{C} + \mathbf{B})(\mathbf{B} - \mathbf{C})y \cdot x = (\mathbf{B}^2 - \mathbf{C}^2)y \cdot x = 0.$$

So $\sqrt{\mathbf{C}}x = 0$ and $\sqrt{\mathbf{B}}x = 0$. Hence, $\mathbf{C}x = \sqrt{\mathbf{C}}\sqrt{\mathbf{C}}x = 0$ and $\mathbf{B}x = 0$. Moreover,

$$\|(\mathbf{B} - \mathbf{C})y\|^2 = (\mathbf{B} - \mathbf{C})^2 y \cdot y = (\mathbf{B} - \mathbf{C})x \cdot y = 0.$$

Thus, $\mathbf{B} = \mathbf{C}$.

Finally, if $\mathbf{A}$ is invertible,

$$\sqrt{\mathbf{A}}\left(\mathbf{A}^{-1}\sqrt{\mathbf{A}}\right) = \left(\mathbf{A}^{-1}\sqrt{\mathbf{A}}\right)\sqrt{\mathbf{A}} = \mathbf{A}^{-1}\sqrt{\mathbf{A}}^2 = I,$$

so $\sqrt{\mathbf{A}}$ is invertible and

$$\sqrt{\mathbf{A}}^{-1} = \mathbf{A}^{-1}\sqrt{\mathbf{A}}. \qquad \square$$

We next show that the functional calculus can be used to give a spectral representation for the operator $\mathbf{A}$.

Let $\mathcal{B} = \mathcal{B}(\sigma(\mathbf{A}))$ be the σ-algebra of Baire subsets of $\sigma(\mathbf{A})$, the same as the σ-algebra of Borel subsets of $\sigma(\mathbf{A})$, (see Remark 1.10). If $f \in C(\sigma)$, let

$$\Psi(f) = f(\mathbf{A}) = \Psi_f.$$

In what follows we use the representation of the dual space of $C(\sigma)$ as the space of all regular Baire signed measures on $\mathcal{B}$; that is, if $F \in C(\sigma)'$, there is a unique regular signed Baire measure μ on $\mathcal{B}$ such that

$$F(f) = \int_\sigma f d\mu$$

for $f \in C(\sigma)$, with $\|F\| = var(\mu)$ ([DS] IV.6.3, [Sw1] 6.5.10).

Theorem 2.7. *For every $\delta \in \mathcal{B}$ there exists a unique orthogonal projection $E(\delta) \in L(H)$ such that*

$$\mu_{x,y}(\delta) = E(\delta)x \cdot y$$

for every $x, y \in H$, where $\mu_{x,y} \in C(\sigma)'$ is the unique regular signed Baire measure on $\mathcal{B}$ satisfying

$$\mu_{x,y}(f) = \int_\sigma f d\mu_{x,y} = \Psi_f x \cdot y.$$

The map $E : \mathcal{B} \to L(H)$ satisfies

    (i) $E(\emptyset) = 0, E(\sigma) = I$,
    (ii) $E(\delta)E(\tau) = E(\delta \cap \tau) = E(\tau)E(\delta)$ for $\delta, \tau \in \mathcal{B}$,
    (iii) $E\left(\bigcup_{j=1}^\infty \delta_j\right)x = \sum_{j=1}^\infty E(\delta_j)x$ for every $x \in H$ and every disjoint
        sequence $\{\delta_j\} \subset \mathcal{B}$.

That is, E is an operator valued measure which is countably additive in the strong operator topology of $L(H)$, such that

$$\Psi_f = f(\mathbf{A}) = \int_\sigma f dE$$

for every $f \in C(\sigma)$. Moreover, $E(\delta)$ commutes with every operator in $L(H)$ which commutes with $\mathbf{A}$.

Proof. For $x, y \in H$ define $\mu_{x,y} \in C(\sigma)'$ by

$$\mu_{x,y}(f) = \Psi_f x \cdot y.$$

Note

$$|\mu_{x,y}(f)| \le \|\Psi_f\| = \|f\| \|x\| \|y\|,$$

so

$$\|\mu_{x,y}\| \le \|x\| \|y\|.$$

For $\delta \in \mathcal{B}$ set

$$b_\delta(x, y) = \mu_{x,y}(\delta).$$

It is easily checked that b_δ is a bounded sesquilinear functional and since Ψ_f is self adjoint, b_δ is symmetric. By Theorem D.4 and Proposition D.6 there exists a self adjoint operator $E(\delta)$ such that

$$b_\delta(x, y) = E(\delta)x \cdot y = \mu_{x,y}(\delta).$$

Now $E(\emptyset) = 0$ and

$$E(\sigma)x \cdot y = \mu_{x,y}(1) = Ix \cdot y,$$

so (i) holds.

To prove (ii), let $f, g \in C(\sigma)$, $x, y \in H$. Then

$$\Psi_f \Psi_g x \cdot y = \Psi_{fg} x \cdot y = \int_\sigma f(t) d(E(t) \Psi_g x \cdot y) = \int_\sigma f(t) g(t) d(E(t) x \cdot y),$$

so

$$E(\delta) \Psi_g x \cdot y = \int_\delta g(t) d(E(t) x \cdot y), \qquad (2.2)$$

for $\delta \in \mathcal{B}$. Also,

$$\mu_{\Psi_g x \cdot y}(f) = \Psi_f \Psi_g x \cdot y = \Psi_f x \cdot \Psi_g y = \mu_{x \cdot \Psi_g y}(f),$$

so

$$\mu_{\Psi_g x \cdot y}(\delta) = \mu_{x \cdot \Psi_g y}(\delta) = E(\delta) \Psi_g x \cdot y$$
$$= \Psi_g E(\delta) x \cdot y = \int_\sigma g(t) d(E(t) E(\delta) x \cdot y). \qquad (2.3)$$

Now, (2.2) and (2.3) imply that
$$E(\delta)E(\tau) = E\left(\delta \bigcap \tau\right)$$
for $\delta, \tau \in \mathcal{B}$, so $E(\delta)E(\tau) = E(\tau)E(\delta)$. In particular,
$$E(\delta) = E\left(\delta \bigcap \delta\right) = E(\delta)E(\delta),$$
which means that $E(\delta)$ is an orthogonal projection.

For (iii), E is clearly finitely additive, so let $\delta_j \downarrow \emptyset$, $\delta_j \in \mathcal{B}$. For $x \in H$,
$$\|E(\delta_j)x\|^2 = E(\delta_j)x \cdot E(\delta_j)x$$
$$= E(\delta_j)x \cdot x = \mu_{x,x}(\delta_j) \downarrow 0,$$
so (iii) holds.

If $f \in C(\sigma)$, the integral $\int_\sigma f dE$ exists by Proposition 1.8, and since
$$\int_\sigma f(t)d(E(t)x \cdot y) = \Psi_f x \cdot y,$$
we have
$$\Psi_f = \int_\sigma f dE.$$

Finally, suppose $\mathbf{T} \in L(H)$ commutes with $\mathbf{A}$. Then, for $x, y \in H$, $f \in C(\sigma)$, by Theorem 2.2, we have
$$\int_\sigma f(t)d(E(t)\mathbf{T}x \cdot y) = \Psi_f \mathbf{T}x \cdot y = \mathbf{T}\Psi_f x \cdot y$$
$$= \Psi_f x \cdot \mathbf{T}^* y = \int_\sigma f(t)d(E(t)x \cdot \mathbf{T}^* y)$$
$$= \int_\sigma f(t)d(\mathbf{T}E(t)x \cdot y),$$
so $E(\delta)\mathbf{T} = \mathbf{T}E(\delta)$ for $\delta \in \mathcal{B}$. $\square$

In the terminology of Definition 1.9, the map $E : \mathcal{B} \to L(H)$ is a spectral measure such that
$$\Psi(f) = \Psi_f = \int_\sigma f dE$$
for every $f \in C(\sigma)$.

According to Proposition 2.8 below, this spectral measure E is unique and it is called the *resolution of the identity* for $\mathbf{A}$.

Proposition 2.8. *Suppose that $F : \mathcal{B} \to L(H)$ is a spectral measure such that*
$$\mathbf{A} = \int_\sigma t dF(t).$$
Then, $E = F$.

Proof. From Theorem 2.7 and Theorem 1.13,

$$\mathbf{A}^n = \int_\sigma t^n dE(t) = \int_\sigma t^n dF(t)$$

for $n = 0, 1, \dots$. Thus,

$$\int_\sigma t^n d(y \cdot E(t)x) = \int_\sigma t^n d(y \cdot F(t)x)$$

for $x, y \in H$, $n = 0, 1, \dots$. So,

$$y \cdot F(\cdot)x = y \cdot E(\cdot)x$$

by the Weierstrass Approximation Theorem. Hence, $E = F$. $\square$

Example 2.9. Let $\mathbf{A} : L^2[0,1] \to L^2[0,1]$ be the multiplication operator $\mathbf{A}f(t) = tf(t)$. Then

$$E(\delta)f = \chi_\delta f$$

is the resolution of the identity for $\mathbf{A}$.

Using the resolution of the identity for the operator $\mathbf{A}$, we may extend the functional calculus $\Psi : C(\sigma) \to L(H)$.

Let $B(\sigma)$ be the space of all real valued bounded Baire functions on σ, equipped with the operations of pointwise sum and product and the sup-norm,

$$\|f\|_\infty = \sup\{|f(t)| : t \in \sigma\}.$$

If $f \in B(\sigma)$, we set

$$\Psi(f) = \int_\sigma f dE.$$

Sometimes we will write $\Psi(f)$ as Ψf to simplify the notation. We then obtain an extension of the functional calculus Ψ to $B(\sigma)$, which from Theorems 2.7 and 1.13 satisfies the following properties:

Theorem 2.10. *The map* $\Psi : B(\sigma) \to L(H)$ *is a continuous algebra homomorphism such that* $\Psi(f)$ *is self adjoint for every* $f \in B(\sigma)$ *and satisfies*

(i) $\|\Psi(f)\| \le \|f\|_\infty$,
(ii) $\|\Psi f(x)\|^2 = \int_\sigma f(t)^2 dE_{xx}(t)$ *for* $f \in B(\sigma)$, $x \in H$,
(iii) *if* $f_j, f \in B(\sigma)$ *are uniformly bounded on* σ *and* $\{f_j\}$ *converges pointwise to* f, *then* $\Psi(f_j)$ *converges to* $\Psi(f)$ *in the strong operator topology of* $L(H)$,

(iv) if $\mathbf{T} \in L(H)$ *commutes with* $\mathbf{A}$, *then* $\mathbf{T}$ *commutes with every* $\Psi(f)$
for $f \in B(\sigma)$.

Given the functional calculus $\Psi : B(\sigma) \to L(H)$ of Theorem 2.10, it is natural to ask what operators in $L(H)$ have the form

$$f(\mathbf{A}) = \Psi(f)$$

for some $f \in B(\sigma)$, that is to say, what is the range of Ψ?

From Theorem 2.10, it follows that any such operator must have the property that it commutes with every operator which commutes with $\mathbf{A}$. We show below that if the space H is separable this necessary condition is also sufficient.

Suppose in what follows that $\mathbf{B} \in L(H)$ is self adjoint and commutes with every operator which commutes with $\mathbf{A}$. We seek a bounded, Baire function $f : \sigma \to \mathbb{R}$ such that $\mathbf{B} = f(\mathbf{A}) = \Psi(f)$.

To motivate the construction below, assume that $\mathbf{B} = f(\mathbf{A})$ for some $f \in B(\sigma)$. Then by Theorem 2.7,

$$\mathbf{B}E(\delta) = E(\delta)\mathbf{B} = \int_\delta f dE,$$

so that f, in some sense, is the "Radon-Nikodym" derivative of $\mathbf{B}E(\cdot)$ with respect to E. To be more precise, let $x \in H$ and note that

$$\mathbf{B}E(\delta)x \cdot x = \int_\delta f d(Ex \cdot x),$$

so $\mathbf{B}E(\cdot)x \cdot x$ is absolutely continuous with respect to $E(\cdot)x \cdot x$. By the Radon Nikodym Theorem,

$$f = \frac{d(\mathbf{B}E(\cdot)x \cdot x)}{d(E(\cdot)x \cdot x)}, \ (E(\cdot)x \cdot x)\text{-a.e.} \tag{2.4}$$

Of course, the difficulty is that the Radon-Nikodym derivative f depends on x. We show below that there is a specific choice of x which will make the construction work.

We begin by establishing two technical lemmas needed later.

Lemma 2.11. *Assume as before that* $\mathbf{B} \in L(H)$ *is self adjoint and commutes with every operator which commutes with* $\mathbf{A}$. *Then, given* $x \in H$ *there exists a sequence of polynomials* $\{p_j\}$ *such that*

$$p_j(\mathbf{A})x \to \mathbf{B}x .$$

Proof. Let M be the closed span of $\{x, \mathbf{A}x, \mathbf{A}^2x, ...\}$ and let P be the orthogonal projection of H onto M. Observe that

$$P\mathbf{A}x = \mathbf{A}x = \mathbf{A}Px,$$

so P commutes with $\mathbf{A}$. Then, $\mathbf{B}$ commutes with P by hypothesis. Therefore,

$$\mathbf{B}x = \mathbf{B}Px = P\mathbf{B}x$$

so $\mathbf{B}x \in M$. Hence, there exist polynomials p_j such that $p_j(\mathbf{A})x \to \mathbf{B}x$. $\square$

Lemma 2.12. *Given $x, y \in H$ there exists a sequence of polynomials $\{p_j\}$ such that*

$$p_j(\mathbf{A})x \to \mathbf{B}x$$

and

$$p_j(\mathbf{A})y \to \mathbf{B}y.$$

Proof. Let $H_1 = H \times H$. Define two bounded self adjoint operators $\mathbf{A}_1$ and $\mathbf{B}_1$ on H_1 by

$$\mathbf{A}_1 = (\mathbf{A}, \mathbf{A}) \text{ and } \mathbf{B}_1 = (\mathbf{B}, \mathbf{B}).$$

Then $\mathbf{A}_1$ and $\mathbf{B}_1$ satisfy the same conditions as $\mathbf{A}$ and $\mathbf{B}$ in Lemma 2.11 so, given $(x, y) \in H_1$, there is a sequence of polynomials $\{p_j\}$ such that

$$p_j(\mathbf{A}_1)(x, y) = (p_j(\mathbf{A})x, p_j(\mathbf{A})y) \to \mathbf{B}_1(x, y) = (\mathbf{B}x, \mathbf{B}y),$$

so

$$p_j(\mathbf{A})x \to \mathbf{B}x \text{ and } p_j(\mathbf{A})y \to \mathbf{B}y. \qquad \square$$

We now fix some notation which will be used below. For $x \in H$ define signed measures β_x and ϵ_x on $\mathcal{B}$ by

$$\beta_x = E(\cdot)\mathbf{B}x \cdot x = \mathbf{B}E(\cdot)x \cdot x\,,$$

$$\epsilon_x = E(\cdot)x \cdot x.$$

Note that β_x is absolutely continuous with respect to ϵ_x. We record some properties of these measures.

Proposition 2.13. *Let $x, y \in H$. Then*

$$\frac{d\beta_x}{d\epsilon_x} = \frac{d\beta_y}{d\epsilon_y}$$

except on the union of 2 sets, one of which is ϵ_x-null and the other which is ϵ_y-null.

Proof. By Lemma 2.12 there is a sequence of polynomials $\{p_j\}$ such that
$$p_j(\mathbf{A})x \to \mathbf{B}x, \text{ and } p_j(\mathbf{A})y \to \mathbf{B}y.$$
For $\delta \in \mathcal{B}$,
$$E(\delta)p_j(\mathbf{A})x \cdot x = p_j(\mathbf{A})E(\delta)x \cdot x = \int_\delta p_j d\epsilon_x$$
so
$$\frac{dE(\cdot)p_j(\mathbf{A})x \cdot x}{d\epsilon_x} = p_j, \ \epsilon_x\text{-a.e.}$$
Similarly,
$$\frac{dE(\cdot)p_j(\mathbf{A})y \cdot y}{d\epsilon_y} = p_j, \ \epsilon_y\text{-a.e.}$$
Now
$$\|E(\delta)p_j(\mathbf{A})x \cdot x - E(\delta)\mathbf{B}x \cdot x\| \le \|p_j(\mathbf{A})x - \mathbf{B}x\| \, \|x\|,$$
so
$$var(E(\cdot)p_j(\mathbf{A}x \cdot x) - \beta_x) \to 0.$$
Therefore,
$$\int_\sigma \left| \frac{dE(\cdot)p_j(\mathbf{A})x \cdot x}{d\epsilon_x} - \frac{d\beta_x}{d\epsilon_x} \right| d\epsilon_x = \int_\sigma \left| p_j - \frac{d\beta_x}{d\epsilon_x} \right| d\epsilon_x \to 0.$$
Similarly,
$$\int_\sigma \left| p_j - \frac{d\beta_y}{d\epsilon_y} \right| d\epsilon_y \to 0.$$
Hence, there is a subsequence $\{p_{n_j}\}$ such that
$$p_{n_j} \to \frac{d\beta_x}{d\epsilon_x}, \ \epsilon_x\text{-a.e.}$$
and
$$p_{n_j} \to \frac{d\beta_y}{d\epsilon_y}, \ \epsilon_y\text{-a.e.}$$
This establishes the result. $\qquad\qquad\qquad\qquad\qquad\qquad\qquad\qquad\Box$

Proposition 2.14. *Let $x \in H$. Then*
$$\frac{d\beta_x}{d\epsilon_x} \in L^\infty(\epsilon_x)$$
with
$$\left| \frac{d\beta_x}{d\epsilon_x} \right| \le \|\mathbf{B}\|, \ \epsilon_x\text{-a.e.}$$

Proof. For $\delta \in \mathcal{B}$,

$$|\beta_x(\delta)| = |\mathbf{B}E(\delta)x \cdot E(\delta)x| \leq \|\mathbf{B}\| \, \|E(\delta)x\|^2$$
$$= \|\mathbf{B}\| \, E(\delta)x \cdot E(\delta)x = \|\mathbf{B}\| \, \epsilon_x(\delta)$$

and the result follows. $\qquad\square$

Proposition 2.15. *The following statements hold:*

    (i) *If $\delta \in \mathcal{B}$ is ϵ_x-null, then δ is $E(\cdot)\mathbf{C}x \cdot \mathbf{C}x$-null for every self adjoint operator $\mathbf{C}$ which commutes with $\mathbf{A}$.*
    (ii) *If $\delta \in \mathcal{B}$ is ϵ_x-null and ϵ_y-null, then δ is ϵ_{x+y}-null.*
    (iii) *If $\delta \in \mathcal{B}$ is ϵ_{x_j}-null and $x_j \to x$, then δ is ϵ_x-null.*

Proof. For (i):

$$E(\delta)\mathbf{C}x \cdot \mathbf{C}x = E(\delta)\mathbf{C}x \cdot E(\delta)\mathbf{C}x = \mathbf{C}E(\delta)x \cdot \mathbf{C}E(\delta)x$$
$$\leq \|\mathbf{C}\|^2 \, \|E(\delta)x\|^2 = \|\mathbf{C}\|^2 \, E(\delta)x \cdot E(\delta)x = \|\mathbf{C}\|^2 \, \epsilon_x(\delta).$$

For (ii):

$$E(\delta)(x+y) \cdot (x+y) = E(\delta)(x+y) \cdot E(\delta)(x+y)$$
$$= \epsilon_x(\delta) + \epsilon_y(\delta) + E(\delta)x \cdot y + y \cdot E(\delta)x$$
$$\leq \epsilon_x(\delta) + \epsilon_y(\delta) + 2\|y\| \, \|E(\delta)x\|$$
$$= \epsilon_x(\delta) + \epsilon_y(\delta) + 2\|y\| \sqrt{\epsilon_x(\delta)}.$$

For (iii):

$$E(\delta)x_j \cdot x_j \to E(\delta)x \cdot x. \qquad\square$$

Assume now that H is separable with $\{x_j\}$ dense in H. Let $\mathcal{P}_j$ be the closed span of $\{x_j, \mathbf{A}x_j, \mathbf{A}^2x_j, ...\}$ and let P_j be the orthogonal projection of H onto $\mathcal{P}_j$. Set

$$y_1 = x_1, \, y_2 = x_2 - P_1x_2, \, ... \, , y_j = x_j - \sum_{i=1}^{j-1} P_i x_j, \, ... \, . \qquad (2.5)$$

Proposition 2.16. *The following statements hold:*

    (i) $P_i \perp P_j$ *or* $\mathcal{R}(P_i) \perp \mathcal{R}(P_j)$, *for* $i \neq j$,
    (ii) $\sum_{i=1}^{\infty} P_i$ *converges pointwise on H to I.*

Proof.   (i): Assume $P_i \perp P_j$ for $i, j < n$. Then, for $i < n$,

$$P_i y_n = P_i x_n - \sum_{k=1}^{n-1} P_i P_k x_n = P_i x_n - P_i x_n = 0.$$

Also,

$$P_i \mathbf{A}^k y_n = \mathbf{A}^k P_i y_n = 0$$

for every k. Hence,

$$P_i(\mathcal{P}_n) = 0 \text{ and } P_i P_n = P_n P_i = 0.$$

Thus, (i) follows by induction.

(ii): Let $x \in H$ and notice that $P_1 + ... + P_n$ is the orthogonal projection of H onto $\mathcal{P}_1 \oplus ... \oplus \mathcal{P}_n$. Since $x_n \in \mathcal{P}_n$ and

$$(P_1 + ... + P_n)x \in \mathcal{P}_1 \oplus ... \oplus \mathcal{P}_{n+p}$$

for every $p \geq 1$,

$$\|x - (P_1 + ... + P_{n+p})x\| \leq \|x - (P_1 + ... + P_n)x\| \leq \|x - x_n\| \,.$$

Since $\{x_k\}$ is dense in H, this implies that $\sum_{i=1}^{\infty} P_i x$ converges to x as desired. $\square$

Referring to the sequence $\{y_j\}$ defined in (2.5), choose $c_k \neq 0$ such that $\sum_{k=1}^{\infty} c_k y_k$ converges absolutely to, say, $y \in H$. We show in what follows that y is a choice of x in (2.4) that makes the construction of f work.

Proposition 2.17. *Let* $x \in H$. *If* $\delta \in \mathcal{B}$ *is* ϵ_y-*null, then* δ *is* ϵ_x-*null.*

Proof.   By (i) in Proposition 2.16,

$$P_j y = \sum_{k=1}^{\infty} P_j c_k y_k = c_j y_j.$$

So,

$$y_j = \frac{1}{c_j} P_j y.$$

By (ii) in Proposition 2.16, δ is ϵ_{y_j}-null. By (i) in Proposition 2.15, δ is also $\epsilon_{p(\mathbf{A})y_j}$-null for every polynomial p.

Now $P_j x \in \mathcal{P}_j$, so δ is $\epsilon_{P_j x}$-null by (i) in Proposition 2.15 and the definition of P_j. Since

$$x = \sum_{j=1}^{\infty} P_j x$$

by (ii) in Proposition 2.16, δ is ϵ_x-null by Proposition 2.15. $\square$

Finally, set

$$f = \frac{d\beta_y}{d\epsilon_y},$$

where by Proposition 2.14, we may assume that $|f(t)| \leq \|\mathbf{B}\|$ for all $t \in \sigma$. From Proposition 2.13,

$$f = \frac{d\beta_x}{d\epsilon_x}, \ \epsilon_x\text{-a.e.}$$

Therefore, we have:

Theorem 2.18. *Assume that the space H is separable and $\mathbf{B}$ is a self adjoint operator in $L(H)$ which commutes with $\mathbf{A}$. Then there exists $f \in B(\sigma)$ such that $\mathbf{B} = f(\mathbf{A})$.*

Proof. For $x \in H$ we have

$$\beta_x(\sigma) = E(\sigma)\mathbf{B}x \cdot x = \mathbf{B}E(\sigma)x \cdot x$$
$$= \mathbf{B}x \cdot x = \int_\sigma \frac{d\beta_x}{d\epsilon_x} d\epsilon_x = \int_\sigma f d\epsilon_x$$

or

$$\mathbf{B} = \int_\sigma f dE.$$ □

Without the separability assumption, Theorem 2.18 is false:

Example 2.19. Let μ be the counting measure on $[0,1]$ and set $H_1 = L^2(\mu) = l^2[0,1]$. Define a self adjoint operator $\mathbf{A}_1$ on H_1 by $\mathbf{A}_1 f(t) = tf(t)$. The resolution of the identity for $\mathbf{A}_1$ is given by

$$E_1(\delta) = \chi_\delta.$$

Let $H_2 = L^2[0,1]$ with respect to the Lebesgue measure and define $\mathbf{A}_2$ on H_2 by $\mathbf{A}_2 f(t) = tf(t)$. The resolution of the identity for $\mathbf{A}_2$ is given by $E_2(\delta) = \chi_\delta$.

Let $H = H_1 \times H_2$ and $\mathbf{A} = (\mathbf{A}_1, \mathbf{A}_2)$. Then $\mathbf{A}$ is self adjoint and its resolution of the identity is given by $E = (E_1, E_2)$.

Consider the operator $\mathbf{P} : H \to H$ given by $\mathbf{P}(u, v) = (u, 0)$. Note that $\mathbf{P}$ is self adjoint and commutes with $\mathbf{A}$.

We claim that $\mathbf{P}$ is not given by $f(\mathbf{A})$ for any bounded Baire function f. Indeed, if there exists such a function f, then f cannot be identically equal to 1. Pick $t \in [0,1]$ such that $f(t) \neq 1$. If $\delta \in \mathcal{B}$,

$$\left\| E(\delta)(\chi_{\{t\}}, 0) \right\|^2 = \left\| (\chi_\delta \chi_{\{t\}}, 0) \right\|^2$$

equals 1 if $t \in \delta$ and equals 0 if $t \notin \delta$.

So,

$$E(\cdot)(\chi_{\{t\}}, 0) \cdot (\chi_{\{t\}}, 0) = \delta_t,$$

the Dirac measure concentrated in t, Example E.3. Therefore,

$$\int_0^1 f d\left(E(\cdot)(\chi_{\{t\}}, 0) \cdot (\chi_{\{t\}}, 0)\right) = f(t) \neq 1,$$

while

$$\mathbf{P}((\chi_{\{t\}}, 0)) \cdot (\chi_{\{t\}}, 0)) = (\chi_{\{t\}}, 0) \cdot (\chi_{\{t\}}, 0) = 1.$$

We can also use the resolution of the identity to construct an extension of the functional calculus in Theorem 2.2, which is an isometric algebra isomorphism instead of just a continuous algebra isomorphism as in Theorem 2.10.

Let E be the resolution of the identity for $\mathbf{A}$. A Baire function $f : \sigma \to \mathbb{R}$ is said to be E-essentially bounded if

$$\inf\{\sup\{|f(t)| : t \in \delta\} : \delta \in \mathcal{B}, E(\delta) = 0\} < \infty.$$

When f is essentially bounded, this infimum is denoted E-essen $\sup(f)$.

Since E is countably additive with respect to the strong operator topology, there exists $\delta_0 \in \mathcal{B}$ such that

$$E(\delta_0) = I,$$

$$E\text{-essen } \sup(f) = \sup\{|f(t)| : t \in \delta_0\}.$$

Thus, there is a bounded Baire function $f_0 : \sigma \to \mathbb{R}$ such that $f_0(t) = f(t)$ for all t except those in a set with E measure 0.

Let $EB(\sigma)$ be the space of all Baire functions $f : \sigma \to \mathbb{R}$ such that E-essen $\sup(f) < \infty$. We say that the elements of $EB(\sigma)$ are essentially bounded in analogy with the situation for the space $L^\infty(\mu)$. We agree to identify functions in $EB(\sigma)$ which are equal except on a set of E measure 0. Thus, the elements of $EB(\sigma)$ are actually equivalence classes of functions, but as usual, we treat the elements of $EB(\sigma)$ as functions.

The space $EB(\sigma)$ is an algebra under the operations of pointwise sum and product and is a Banach space under the norm

$$\|f\| = E\text{-essen } \sup(f).$$

If $f \in EB(\sigma)$, we may define the integral of f with respect to E by choosing any bounded Baire function f_0 which is equal to f except for a set of E measure 0 and setting

$$\int_\sigma f dE = \int_\sigma f_0 dE.$$

By Corollary 1.19 the value of the integral is independent of the choice of f_0.

We have the following extension of the functional calculus in Theorem 2.2 to an isometric algebra isomorphism on $EB(\sigma)$:

Theorem 2.20. *The map* $\Psi : EB(\sigma) \to L(H)$ *defined by*

$$\Psi(f) = \int_\sigma f dE$$

is an isometric algebra isomorphism which extends the map Ψ of Theorem 2.2 and satisfies:

(i) $\|\Psi f\|^2 = \int_\sigma f^2(t) d(E(t)x \cdot x)$ *for* $f \in EB(\sigma)$, $x \in H$,

(ii) *if $\{f_j\}$ is a bounded sequence in $EB(\sigma)$, $f \in EB(\sigma)$ and if $\{f_j\}$ converges to f except on a set with E measure 0, then $\Psi(f_j) \to \Psi(f)$ in the strong operator topology of $L(H)$.*

Proof. The map Ψ is obviously linear and Ψ is an algebra homomorphism by the argument in Theorem 2.2. Statements (i) and (ii) follow from Theorem 2.2.

Let $f \in EB(\sigma)$ and suppose $\delta \in \mathcal{B}$ and $E(\delta) = I$. Then

$$\|\Psi(f)\| = \left\| \int_\delta f dE \right\| \leq \sup\{|f(t)| : t \in \delta\},$$

so

$$\|\Psi f\| \leq E\text{-essen sup}(f).$$

Conversely, suppose $r < E\text{-essen sup}(f)$ and put

$$\delta_r = \{t : |f(t)| > r\}$$

so that $E(\delta_r) \neq 0$. Pick $x \in H$ such that $E(\delta_r)x = x \neq 0$. By (i),

$$\|\Psi(f)\|^2 = \int_\sigma f^2(t) d(E(t)E(\delta_r)x \cdot x)$$

$$= \int_\sigma f^2(t) d(E(t \cap \delta_r)x \cdot x) = \int_{\delta_r} f^2(t) d(E(t)x \cdot x) \geq r^2 \|x\|^2.$$

Since $x \neq 0$, this implies $\|\Psi(f)\| \geq r$, so

$$\|\Psi(f)\| \geq E\text{-essen sup}(f).$$

Hence,

$$\|\Psi(f)\| = E\text{-essen sup}(f)$$

and Ψ is an isometry. $\qquad \square$

A problem associated with self adjoint operators is the simultaneous representation of a family of operators by means of a single spectral measure. We give a simple example of such a representation which can be derived from the functional calculus of Theorem 2.2.

Definition 2.21. Let X be a Banach space. A family

$$\{\mathbf{T}_t : t \in [0, \infty)\} \subset L(X)$$

is said to be a uniformly continuous semi-group if

  (a) $\mathbf{T}_{s+t} = \mathbf{T}_s \mathbf{T}_t$ for $s, t \geq 0$,
  (b) $\mathbf{T}_0 = I$,
  (c) the map $t \to \mathbf{T}_t$ from $[0, \infty) \to L(X)$ is continuous with respect to the uniform operator topology.

The semigroup is called strongly continuous when the map in (c) is continuous with respect to the strong operator topology.

Example 2.22. Let $\mathbf{T} \in L(X)$ and set

$$\mathbf{T}_t = \exp(t\mathbf{T}) = e^{t\mathbf{T}}.$$

Then $\{\mathbf{T}_t : t \geq 0\}$ defines a uniformly continuous semi-group.

Example 2.23. Let $X = L^2(\mathbb{R})$ with respect to the Lebesgue measure. For $f \in L^2(\mathbb{R})$ and $t \in [0, \infty)$, set

$$\tau_t f(s) = f(s + t).$$

Then $\mathbf{T}_t f = \tau_t f$ defines a strongly continuous semi-group which is not uniformly continuous.

For strongly continuous semi-groups of self adjoint operators, we have the following interesting representation theorem:

Theorem 2.24. *Let $\{\mathbf{T}_t : t \in [0, \infty)\}$ be a strongly continous semi-group of self adjoint operators in $L(H)$. Then there exists a compact subset $K \subset [0, \infty)$ and a spectral measure $E : \mathcal{B}(K) \to L(H)$ such that*

$$\mathbf{T}_t = \int_K s^t dE(s).$$

Proof. Since $\mathbf{T}_t = (\mathbf{T}_{t/2})^2$, each $\mathbf{T}_t$ is a positive operator, so the spectrum of $\mathbf{T}_1$ is a compact set K contained in $[0, \infty)$. Let E be the resolution of the identity for $\mathbf{T}_1$. According to Theorem 1.13, define a strongly continuous semi-group of self adjoint operators in $L(H)$ by

$$\mathbf{B}_t = \int_K s^t dE(s).$$

Now $\mathbf{T}_1 = \mathbf{B}_1$ and moreover, $\mathbf{T}_t = \mathbf{B}_t$ for $t = 1/2, ..., 1/2^k$, since by Theorem 2.6, each positive operator has a unique square root.

By (a) in Definition 2.21, $\mathbf{T}_t = \mathbf{B}_t$ for $t = j/2^k$, $j, k \in \mathbb{N}$. By continuity of the maps $t \to \mathbf{T}_t$ and $t \to \mathbf{B}_t$, it follows that $\mathbf{T}_t = \mathbf{B}_t$ for $t \in [0, \infty)$. $\square$

Remark 2.25. The spectral representation of a bounded self adjoint operator given in Theorem 2.7 is one of the most important theorems in operator theory. The theorem was originally established by Hilbert and since then many proofs have been given. See [RN] p.275, for some historic remarks and other proofs. Riesz and Nagy give a spectral representation of a self adjoint operator as a Stieltjes type integral over an interval containing the spectrum; see also [BN] and [TL]. Steen ([Ste]) has also given a sketch of the history of spectral theory.

The literature of the theory of semi-groups of operators is huge and has significant applications to differential equations. For further information, see [HP], [DS] and [Da].

Chapter 3

Functions of Several Commuting Self Adjoint Operators

Suppose that $\mathbf{S}_1, ..., \mathbf{S}_n$ are commuting, bounded, self adjoint operators defined on a complex Hilbert space H. If $p(s_1, ..., s_n)$ is a real polynomial in n variables, we have a natural definition for $p(\mathbf{S}_1, ..., \mathbf{S}_n) \in L(H)$. If $\mathcal{P}$ is the algebra of all real polynomials in n variables (with the usual pointwise operations of sums and products), we have a straightforward functional calculus $\Psi : \mathcal{P} \rightarrow L(H)$, $\Psi p = p(\mathbf{S}_1, ..., \mathbf{S}_n)$, which is an algebra isomorphism. If $\sigma_i = \sigma(\mathbf{S}_i)$ and if $C(\sigma_1 \times ... \times \sigma_n)$ is the algebra of all real valued, continuous functions on $\sigma_1 \times ... \times \sigma_n$ with the sup-norm, it is natural to ask if it is possible to extend the functional calculus Ψ to $C(\sigma_1 \times ... \times \sigma_n)$ as was done in Chapter 2, i.e., can we define $f(\mathbf{S}_1, ..., \mathbf{S}_n)$ for any continuous, real valued function on $\sigma_1 \times ... \times \sigma_n$? In order to obtain such an extension in Chapter 2, we used the equality

$$\|p(\mathbf{A})\| = \sup\{|p(t)| : t \in \sigma(\mathbf{A})\}, \tag{3.1}$$

where $\mathbf{A} \in L(H)$ was self adjoint and p was a real polynomial in one variable. It is natural to ask if the analogue of (3.1) holds for polynomials in several variables. That is, does the equality

$$\|p(\mathbf{S}_1, ..., \mathbf{S}_n)\| = \sup\{|p(s_1, ..., s_n)| : s_i \in \sigma_i, i = 1, ..., n\} \tag{3.2}$$

hold? We show below in Example 3.1 that this is not the case. However, the inequality

$$\|p(\mathbf{A})\| \le \sup\{|p(t)| : t \in \sigma(\mathbf{A})\}$$

would have sufficed to obtain the extension of the functional calculus Ψ in Chapter 2 so it is natural to ask if the analogue of this inequality holds in the n dimensional case. That is, does the inequality

$$\|p(\mathbf{S}_1, ..., \mathbf{S}_n)\| \le \sup\{|p(s_1, ..., s_n)| : s_i \in \sigma_i, i = 1, ..., n\} \tag{3.3}$$

hold for $p \in \mathcal{P}$? We show below that (3.3) does indeed hold and can be used to extend the functional calculus Ψ from $C(\sigma_1 \times ... \times \sigma_n)$ to $L(H)$.

We begin by giving an example which shows that, in general, the equality (3.1) does not hold for polynomials in two variables.

Example 3.1. Let $H = L^2[0,1]$ (Lebesgue measure). Define $\mathbf{S}, \mathbf{T} \in L(H)$ by $\mathbf{S}f(t) = tf(t)$ and $\mathbf{T}f(t) = (1-t)f(t)$; note $\mathbf{S}, \mathbf{T}$ are self adjoint and commute. Then $\sigma(\mathbf{S}) = \sigma(\mathbf{T}) = [0,1]$. Define a polynomial p by $p(u,v) = u + v$. Then $p(\mathbf{S}, \mathbf{T}) = I$ so $\|p(\mathbf{S}, \mathbf{T})\| = 1$, $\sigma(p(\mathbf{S}, \mathbf{T})) = \{1\}$, $p(\sigma(\mathbf{S}), \sigma(\mathbf{T})) = [0,2]$, $\|p\|_\infty = \sup\{|p(u,v)| : u,v \in [0,1]\} = 2$.

Thus, (3.1) fails in this example. This example also shows that there is no analogue of the Spectral Mapping Theorem (2.4) for polynomials in two variables.

We show that the inequality (3.3) holds in the case of two variables by constructing the product of the resolutions of the identity for two commuting, self adjoint operators. We will show how to construct the product of two general spectral measures and then apply the result to the case where the spectral measures are the resolutions of the identities for commuting self adjoint operators. The general case can then be obtained by induction.

Let Ω and Λ be compact, Hausdorff spaces and let U and V be commuting spectral measures on $\mathcal{B}(\Omega)$ and $\mathcal{B}(\Lambda)$, respectively. We define the product of the spectral measures U and V and use this product to obtain the desired inequality (3.3).

Let $\mathcal{R}$ be the semi-algebra of all measurable rectangles $\mathbf{A} \times \mathbf{B}$, $\mathbf{A} \in \mathcal{B}(\Omega)$, $\mathbf{B} \in \mathcal{B}(\Lambda)$. Let $\mathcal{A}$ be the algebra generated by $\mathcal{R}$ and let Σ be the σ-algebra generated by $\mathcal{A}$. For $x \in H$ let $U_x : \mathcal{B}(\Omega) \to \mathcal{H}$ be the vector valued measure defined by $U_x(\mathbf{A}) = U(\mathbf{A})x$; V_x is defined similarly. For $x, y \in H$ we define the product of U_y and V_x on $\mathcal{R}$ by $U_y \cdot V_x(\mathbf{A} \times \mathbf{B}) = U_y(\mathbf{A}) \cdot V_x(\mathbf{B})$. By the usual (tedious!) calculations one obtains that $U_y \cdot V_x$ is finitely additive on $\mathcal{R}$ and, therefore, has a unique, finitely additive extension, $U_y \cdot V_x$, to $\mathcal{A}$. Moreover, we have:

Proposition 3.2. *The finitely additive extension* $U_y \cdot V_x : \mathcal{A} \to \mathbb{C}$ *is bounded with* $|U_y \cdot V_x(C)| \leq \|y\| \|x\|$ *for all* $C \in \mathcal{A}$.

Proof.   If $\mathbf{A} \in \mathcal{B}(\Omega)$, $\mathbf{B} \in \mathcal{B}(\Lambda)$, we have

$$|U_y \cdot V_x(\mathbf{A} \times \mathbf{B})| \leq \|U(\mathbf{A})y\| \|V(\mathbf{B})x\| .$$

If $C \in \mathcal{A}$ and $C = \cup_{i=1}^n \mathbf{A}_i \times \mathbf{B}_i$ with $\{\mathbf{A}_i \times \mathbf{B}_i : i = 1, ..., n\} \subset \mathcal{R}$

pairwise disjoint, then

$$|U_y \cdot V_x(C)| = \left| \sum_{i=1}^{n} U_y \cdot V_x(\mathbf{A}_i \times \mathbf{B}_i) \right|$$

$$= \left| \sum_{i=1}^{n} y \cdot U(\mathbf{A}_i)V(\mathbf{B}_i)x \right| = \left| y \cdot \sum_{i=1}^{n} U(\mathbf{A}_i)V(\mathbf{B}_i)x \right| \le \|y\| \, \|x\|$$

since $\sum_{i=1}^{n} U(\mathbf{A}_i)V(\mathbf{B}_i)$ is an orthogonal projection (this uses commutativity and the properties in Definition 1.9). □

We seek to extend $U_y \cdot V_x$ to Σ. For this we need the countable additivity of $U_y \cdot V_x$ on $\mathcal{A}$ (Lemma 1.20). We establish this by using Theorem 1.28 on products of measures.

Proposition 3.3. *$U_y \cdot V_x$ is countably additive on $\mathcal{A}$.*

Proof. Proposition 3.2 shows that condition (a) of Theorem 1.28 holds. Since $U_y \cdot V_x(\mathbf{A} \times \mathbf{B}) = U_y(\mathbf{A}) \cdot V_x(\mathbf{B})$ for $\mathbf{A} \in \mathcal{B}(\Omega)$, $\mathbf{B} \in \mathcal{B}(\Lambda)$, conditions (b) and (c) of Theorem 1.28 also hold, so the result follows from this theorem. □

Corollary 3.4. *$U_y \cdot V_x$ has a unique, countably additive extension to Σ satisfying $|U_y \cdot V_x(\mathbf{C})| \le \|y\| \, \|x\|$ for $\mathbf{C} \in \Sigma$.*

Proof. Proposition 3.3 and Lemma 1.20. □

We will require a simple Fubini Theorem with respect to the product measure $U_y \cdot V_x$.

Theorem 3.5. *(Fubini) Let $f : \Omega \to \mathbb{C}$, $g : \Lambda \to \mathbb{C}$ be bounded, Baire functions and define $f \otimes g : \Omega \times \Lambda \to \mathbb{C}$ by $f \otimes g(s, t) = f(s)g(t)$. Then*

$$\int_{\Omega \times \Lambda} f \otimes g \, dU_y \cdot V_x = \int_{\Omega} f \, dU_y \cdot \int_{\Lambda} g \, dV_x.$$

Proof. First suppose $f = \sum_{i=1}^{m} a_i \chi_{\mathbf{A}_i}$ and $g = \sum_{j=1}^{n} b_j \chi_{\mathbf{B}_j}$ are Baire, simple functions. Then

$$\int_{\Omega \times \Lambda} f \otimes g \, dU_y \cdot V_x = \sum_{i=1}^{m} \sum_{j=1}^{n} a_i b_j U_y(\mathbf{A}_i) \cdot V_x(\mathbf{B}_j)$$

$$= \sum_{i=1}^{m} a_i U_y(\mathbf{A}_i) \cdot \sum_{j=1}^{n} b_j V_x(\mathbf{B}_j) = \int_{\Omega} f \, dU_y \cdot \int_{\Lambda} g \, dV_x.$$

For the general case pick a sequence $\{f_k\}$ ($\{g_k\}$) of simple, Baire functions converging uniformly to f (g). Then $\{f_k \otimes g_k\}$ converges uniformly to $f \otimes g$ so

$$\int_{\Omega \times \Lambda} f \otimes g \, dU_y \cdot V_x = \lim \int_{\Omega \times \Lambda} f_k \otimes g_k dU_y \cdot V_x$$

$$= \lim \int_\Omega f_k dU_y \cdot \int_\Lambda g_k dV_x = \int_\Omega f dU_y \cdot \int_\Lambda g dV_x. \qquad \square$$

We can now employ the results above to construct the product $U \times V$ of the spectral measures U and V. We begin by defining the product $U \times V$ (with respect to the product in $L(H)$) on $\mathcal{R}$ by setting $U \times V(\mathbf{A} \times \mathbf{B}) = U(\mathbf{A})V(\mathbf{B})$, where $\mathbf{A} \in \mathcal{B}(\Omega)$, $\mathbf{B} \in \mathcal{B}(\Lambda)$. As before $U \times V$ is finitely additive on $\mathcal{R}$ and has a unique, finitely additive extension to $\mathcal{A}$.

Theorem 3.6. *$U \times V$ has a unique extension to Σ which is countably additive with respect to the strong operator topology of $L(H)$ and satisfies $\|U \times V(C)\| \leq 1$ for $C \in \Sigma$. Moreover, $U \times V$ is a spectral measure.*

Proof.   Note that if $x, y \in H$, $\mathbf{A} \in \mathcal{B}(\Omega)$, $\mathbf{B} \in \mathbf{B}(\Lambda)$, then $U_y \cdot V_x(\mathbf{A} \times \mathbf{B}) = y \cdot U \times V(\mathbf{A} \times \mathbf{B})x$ so from uniqueness it follows that $U_y \cdot V_x = y \cdot (U \times V)x$ on $\mathcal{A}$. From Propositions 3.2 and 3.3 and Theorem 1.22, it follows that $U \times V$ has a unique extension to Σ which is countably additive with respect to the strong operator topology and satisfies $|y \cdot U \times V(C)x| \leq \|y\| \|x\|$ for all $x, y \in H$, $C \in \Sigma$. Hence, $\|U \times V\| \leq 1$ for $C \in \Sigma$.

We now check condition (iv) of Definition 1.9. It is readily checked that $U \times V(C \cap D) = U \times V(C)U \times V(D)$ holds for $C, D \in \mathcal{R}$ and then for $C, D \in \mathcal{A}$.

Fix $C \in \mathcal{A}$ and put

$$\mathcal{D} = \{D \in \Sigma : U \times V(C \cap D) = U \times V(C)U \times V(D)\}.$$

From the observation above, $\mathcal{D} \supset \mathcal{A}$. We claim that $\mathcal{D}$ is a monotone class. Suppose $D_i \in \mathcal{D}$ and $D_i \uparrow D$. Then $C \cap D_i \uparrow C \cap D$ so by the countable additivity of $U \times V$, it follows that $U \times V(C \cap D) = U \times V(C)U \times V(D)$ so $D \in \mathcal{D}$. Decreasing sequences are treated similarly. Thus, $\mathcal{D}$ is a monotone class and from the Monotone Class Lemma, $\mathcal{D} = \Sigma$.

Now fix $D \in \Sigma$ and put

$$\mathcal{C} = \{C \in \Sigma : U \times V(C \cap D) = U \times V(C)U \times V(D)\}.$$

From above, we have $\mathcal{C} \supset \mathcal{A}$, and, as above, $\mathcal{C} = \Sigma$. That is, condition (d) of Definition 1.9 holds.

For condition (b) of Definition 1.9, note that $U \times V(C)$ is a projection from the intersection property established above. Therefore, it remains to show that $U \times V(C)$ is self adjoint. First, $U \times V(C)$ is self adjoint for $C \in \mathcal{R}$ by commutativity and then self adjoint for $C \in \mathcal{A}$ by commutativity. Put

$$\mathcal{E} = \{C \in \Sigma : U \times V(C) \text{ is self adjoint}\}.$$

Then $\mathcal{A} \subset \mathcal{E}$ by what we have just observed. We claim that $\mathcal{E}$ is a monotone class. Suppose $C_i \uparrow C$. If $x, y \in H$, by countable additivity,

$$y \cdot U \times V(C_i)x = U \times V(C_i)y \cdot x \to y \cdot U \times V(C)x = U \times V(C)y \cdot x$$

so $U \times V(C)$ is symmetric and $C \in \mathcal{E}$. Decreasing sequences are treated similarly so $\mathcal{E} = \Sigma$ by the Monotone Class Lemma. $\qquad\square$

From Theorem 3.6 we have another version of Fubini's Theorem.

Theorem 3.7. *(Fubini) Let $f : \Omega \to \mathbb{C}$ and $g : \Lambda \to \mathbb{C}$ be bounded, Baire functions. Then $\int_{\Omega \times \Lambda} f \otimes g \, dU \times V = \int_{\Omega} f dU \int_{\Lambda} g dV$.*

Proof. Note all three integrals exist (Proposition 1.8). The equality follows from Theorem 3.5. $\qquad\square$

We now have the machinery to establish inequality (3.3). Let $\mathbf{S}, \mathbf{T} \in L(H)$ be commuting, self adjoint operators with resolutions of the identity E and F, respectively. Let $\sigma = \sigma(\mathbf{S})$ and $\tau = \sigma(\mathbf{T})$. From Fubini's Theorem we have:

Theorem 3.8. *If p is a real polynomial in two variables, then*

$$p(\mathbf{S}, \mathbf{T}) = \int_{\sigma \times \tau} p(s, t) dE \times F(s, t).$$

Proof. Since $\mathbf{S}^i = \int_{\sigma} s^i dE(s)$ and $\mathbf{T}^j = \int_{\tau} t^j dF(t)$ for $i, j \geq 0$, we have from Theorem 3.7 that $\mathbf{S}^i \mathbf{T}^j = \int_{\sigma \times \tau} s^i t^j dE \times F(s, t)$ so the result follows from linearity. $\qquad\square$

Corollary 3.9. *If p is a real polynomial in two variables, then*

$$\|p(\mathbf{S}, \mathbf{T})\| \leq \sup\{|p(s, t)| : s \in \sigma, t \in \tau\} = \|p\|_{\infty}.$$

Proof. Since $E \times F$ is a spectral measure (Theorem 3.6), the result follows from Theorem 1.13. $\qquad\square$

Let $\mathcal{P}$ be the algebra of all real polynomials with the usual operations of pointwise sum and product and equip $\mathcal{P}$ with the sup norm,

$$\sup\{|p(s,t)| : s \in \sigma, t \in \tau\} = \|p\|_\infty .$$

Define $\Psi : \mathcal{P} \to L(H)$ by $\Psi p = p(\mathbf{S},\mathbf{T})$. Then Ψ is an algebra isomorphism and from Corollary 3.9 Ψ is continuous with $\|\Psi p\| \le \|p\|_\infty$. If $C(\sigma \times \tau)$ is the algebra of continuous, real valued functions defined on $\sigma \times \tau$ equipped with the sup-norm, it follows from the Weierstrass Approximation Theorem that Ψ has a unique continuous extension

$$\Psi : C(\sigma \times \tau) \to L(H)$$

which is an algebra isomorphism satisfying $\|\Psi f\| \le \|f\|_\infty$ for $f \in C(\sigma \times \tau)$. If $f \in C(\sigma \times \tau)$, we write $\Psi f = f(\mathbf{S},\mathbf{T})$ and we have from Theorem 3.8 the integral representation,

$$f(\mathbf{S},\mathbf{T}) = \int_{\sigma \times \tau} f(s,t)dE \times F(s,t).$$

Using induction we can extend the result above to the case of polynomials and functions of several variables. To see that this is the case, let $\mathbf{R} \in L(H)$ be self adjoint and commute with $\mathbf{S}$ and $\mathbf{T}$ with $\rho = \sigma(\mathbf{R})$ and G the resolution of the identity for $\mathbf{R}$. From Theorem 3.7 we obtain

$$\mathbf{R}^i \mathbf{S}^j \mathbf{T}^k = \int_\rho r^i dG(r) \int_\sigma s^j dE(s) \int_\tau t^k dF(t)$$

$$= \int_\rho r^i dG(r) \int_{\sigma \times \tau} s^j t^k dE \times F(s,t).$$

But, $E \times F$ is a spectral measure (Theorem 3.6) so another application of Theorem 3.7 gives

$$\mathbf{R}^i \mathbf{S}^j \mathbf{T}^k = \int_{\rho \times \sigma \times \tau} r^i s^j t^k dG \times (E \times F)(r,s,t).$$

Thus, if p is a real polynomial in three variables,

$$p(\mathbf{R},\mathbf{S},\mathbf{T}) = \int_{\rho \times \sigma \times \tau} p(r,s,t)dG \times E \times F(r,s,t)$$

and the construction above can be carried out for continuous functions of 3 variables.

We can use the product measure $E \times F$ to further extend the functional calculus Ψ to the space $B(\sigma \times \tau)$ of bounded, complex valued, Baire functions on $\sigma \times \tau$. That is, we define

$$\Psi f = f(\mathbf{S},\mathbf{T}) = \int_{\sigma \times \tau} f(s,t)dE \times F(s,t).$$

From Theorem 1.13, $\Psi : B(\sigma \times \tau) \to L(H)$ is an algebra isomorphism which extends Ψ from $C(\sigma \times \tau)$ and satisfies $\|\Psi f\| \le \|f\|_\infty$ for $f \in B(\sigma \times \tau)$. The case of functions of several variables is treated similarly.

Chapter 4

The Spectral Theorem for Normal Operators

In this chapter we extend the spectral theorem for self adjoint operators given in Chapter 2 to normal operators. We give two proofs of the spectral theorem for normal operators. The first proof is based on the product of the resolution of the identities for the two self adjoint components of a normal operator and the second proof is based on an extension of the functional calculus developed in Chapter 2 for self adjoint operators. We also establish a spectral theorem for unitary operators.

Suppose $\mathbf{N}$ is a normal operator on the complex Hilbert space H with $\mathbf{N} = \mathbf{S} + i\mathbf{T}$, where $\mathbf{S}$ and $\mathbf{T}$ are commuting, self adjoint, operators. Let $\sigma = \sigma(\mathbf{S})$ and $\tau = \sigma(\mathbf{T})$ and E, F be the resolutions of the identity for $\mathbf{S}$ and T, respectively. Since $\mathbf{S}$ and $\mathbf{T}$ commute, E and F commute (Theorems 1.13 and 2.10). Therefore, the product $E \times F : \mathcal{B}(\sigma \times \tau) \to L(H)$ exists and is a spectral measure (Theorem 3.6). Moreover, we have the following spectral representation theorem.

Theorem 4.1.

(i) $\mathbf{N} = \int_{\sigma \times \tau} (x + iy) dE \times F(x, y)$,

(ii) $\mathbf{N}^* = \int_{\sigma \times \tau} (x - iy) dE \times F(x, y)$.

Proof. From Theorems 2.10 and 3.7,

$$\int_{\sigma \times \tau} (x + iy) dE \times F(x, y) = \int_{\sigma} x dE(x) \int_{\tau} dF(y) + i \int_{\sigma} dE(x) \int_{\tau} y dF(y)$$

$$= \mathbf{S} + i\mathbf{T} = \mathbf{N}.$$

(ii) is similar. $\qquad\qquad\square$

If $z = x + iy \in \sigma \times \tau$ and $f : \sigma \times \tau \to \mathbb{C}$ is a bounded, Baire function and if we write

$$\int_{\sigma \times \tau} f(z) dE \times F(z) = \int_{\sigma \times \tau} f(x + iy) dE \times F(x, y),$$

we then have

$$\mathbf{N} = \int_{\sigma \times \tau} z dE \times F(z), \quad \mathbf{N}^* = \int_{\sigma \times \tau} \overline{z} dE \times F(z).$$

As before we can now define a functional calculus by setting

$$f(\mathbf{N}) = \int_{\sigma \times \tau} f(z) dE \times F(z).$$

Consider this functional calculus for polynomials. Suppose p is a complex polynomial of the form

$$p(z, \overline{z}) = \sum a_{ij} z^i \overline{z}^j.$$

From the formulas above and Theorem 1.13, we obtain

$$\int_{\sigma \times \tau} p(z, \overline{z}) dE \times F(z) = \sum a_{ij} \mathbf{N}^i (\mathbf{N}^*)^j, \tag{4.1}$$

a very different transformation than the one previously obtained.

The spectral measure for normal operators given in Theorem 4.1 is defined on the Baire sets of $\sigma \times \tau$ which contains the spectrum of $\mathbf{N}$ whereas the resolution of the identity for a self adjoint operator $\mathbf{A}$ as given in Theorem 2.10 is defined on the spectrum $\sigma(\mathbf{A})$ of $\mathbf{A}$. We show that the support of the product measure $E \times F$ is actually $\sigma(\mathbf{N})$ so the resolution of the identity for $\mathbf{N}$ can be taken to be a spectral measure on $\sigma(\mathbf{N})$.

Theorem 4.2. $\sigma(\mathbf{N}) = supp(E \times F)$.

Proof.   For convenience, let $\Omega = \sigma \times \tau$ and $G = E \times F$. Let $\lambda \in supp(G)$. Let $\epsilon > 0$ and set

$$U = \{z \in \mathbb{C} : |\lambda - z| < \epsilon\}.$$

Then $G(U) \neq 0$ since $\lambda \in supp(G)$. Pick $x \in H$ such that $G(U)x = x, \|x\| \leq 1$. Then

$$G_{xx}(\Omega \backslash U) = x \cdot x - G(U)x \cdot x = 0.$$

By Theorem 1.13,

$$\epsilon^2 = \epsilon^2 G_{xx}(U) \geq \int_U |\lambda - z|^2 dG_{xx}(z) = \int_\Omega |\lambda - z|^2 dG_{xx}(z) = \|(\mathbf{N} - \lambda)x\|^2.$$

By Theorem C.14 $\lambda \in \sigma(\mathbf{N})$ so $supp(G) \subset \sigma(\mathbf{N})$.

Suppose $\lambda \notin supp(G)$. Then there exists an open set U containing λ such that $G(U) = 0$. Define $h(z) = 1/(\lambda - z)$ if $z \notin U$ and $h(z) = 0$ if $z \in U$. Then h is a bounded, Baire function and by the properties of the functional calculus

$$h(\mathbf{N})(\lambda - \mathbf{N}) = \int_\Omega h(z)(\lambda - z) dG(z) = \int_{\Omega \backslash U} dG = \int_\Omega dG = I.$$

Hence, $\lambda - \mathbf{N}$ is invertible and $\lambda \notin \sigma(\mathbf{N})$. Thus, $\sigma(\mathbf{N}) \subset supp(G)$. $\square$

We now have a sharper form of the spectral theorem for normal operators.

Theorem 4.3. *There is a unique spectral measure G defined on $\mathcal{B}(\sigma(\mathbf{N}))$ such that*

$$\mathbf{N} = \int_{\sigma(\mathbf{N})} z \, dG(z).$$

Moreover, (4.1) holds with $\sigma \times \tau$ replaced by $\sigma(\mathbf{N})$.

Proof. We merely restrict $E \times F$ to $\mathcal{B}(\sigma(\mathbf{N}))$ and cite Corollary 1.19. The uniqueness follows as in Proposition 2.8. □

A similar approach to the spectral theorem for normal operators is given in [Be2].

We now give another proof of the spectral theorem for normal operators which is analogous to the proof of the spectral theorem for self adjoint operators given in Chapter 2.2. The proof of the spectral theorem for a self adjoint operator $\mathbf{A}$ given in Theorem 2.2 was based on the functional calculus Ψ which was an isometric isomorphism from $C_{\mathbb{R}}(\sigma(\mathbf{A}))$ onto the closed subalgebra $\mathcal{A}$ of $L(H)$ generated by $\mathbf{A}$ and the identity operator I which carries a real polynomial p to the operator $p(\mathbf{A})$. We consider what form the analogue of this functional calculus should take for a normal operator.

The first difference between the self adjoint case and the case for normal operators is that the spectrum of a normal operator is, in general, a compact subset of $\mathbb{C}$ so the identity polynomial $p(z) = z$ which should be associated with the normal operator $\mathbf{N}$ is complex valued so we must consider complex polynomials and then the space $C_{\mathbb{C}}(\sigma(\mathbf{N}))$ of complex valued, continuous functions on $\sigma(\mathbf{N})$ instead of $C_{\mathbb{R}}(\sigma(\mathbf{N}))$. However, the complex polynomials are not, in general, dense in $C_{\mathbb{C}}(\sigma(\mathbf{N}))$ [consider the function $f(z) = \overline{z}$] so we cannot simply map the polynomial p to the operator $p(\mathbf{N})$ and extend this map to $C_{\mathbb{C}}(\sigma(\mathbf{N}))$ as was done in the case of self adjoint operators. Recall from the Stone-Weierstrass Theorem that a subalgebra is dense in $C_{\mathbb{C}}(\sigma(\mathbf{N}))$ if it contains the constant functions, separates the points of $\sigma(\mathbf{N})$ and is closed under conjugation. If $p(z) = z = x + iy$ is the identity polynomial and is associated with the operator $\mathbf{N}$, then the conjugate polynomial $\overline{p}(z) = \overline{z} = x - iy$ should be associated with the operator $\mathbf{N}^*$. This suggests that we should consider polynomials of the form

$$p(z, \overline{z}) = \sum a_{ij} z^i \overline{z}^j, \, a_{ij} \in \mathbb{C},$$

and associate such a polynomial with the operator

$$p(\mathbf{N}, \mathbf{N}^*) = \sum a_{ij} \mathbf{N}^i (\mathbf{N}^*)^j$$

(see also (4.1)). We then need to show that the map Ψ which sends the polynomial $p(z, \overline{z})$ to the operator $p(\mathbf{N}, \mathbf{N}^*)$ is an isometric algebra isomorphism from the space $\mathcal{P}$ of polynomials of the form $p(z, \overline{z})$ into $L(H)$.

We now give the details of such a construction; this construction is due to Whitley ([Wh]). Let $\mathcal{P}$ be the space of all complex polynomials of the form $p(z, \overline{z}) = \sum a_{ij} z^i \overline{z}^j, a_{ij} \in \mathbb{C}$, with the sup norm

$$\|p\|_\infty = \sup\{|p(z, \overline{z})| : z \in \sigma(\mathbf{N})\}.$$

Then $\mathcal{P}$ is an algebra under the usual pointwise operations and since $\mathcal{P}$ separates the points of $\sigma(\mathbf{N})$ and is closed under conjugation $\mathcal{P}$ is dense in $C_{\mathbb{C}}(\sigma(\mathbf{N}))$. Let $\mathcal{A}$ be the closed subalgebra of $L(H)$ generated by $\mathbf{N}, \mathbf{N}^*$ and the identity operator I and define the map $\Psi : \mathcal{P} \to \mathcal{A}$ by

$$\Psi : p(z, \overline{z}) \to p(\mathbf{N}, \mathbf{N}^*)$$

so that $\Psi\mathcal{P}$ is dense in $\mathcal{A}$. It is readily checked that Ψ is an algebra isomorphism. We now show that Ψ is an isometry, i.e.,

$$\|p(\mathbf{N}, \mathbf{N}^*)\| = \|p\|_\infty,$$

and then we can extend Ψ to be an isometric isomorphism from $C_{\mathbb{C}}(\sigma(\mathbf{N}))$ onto $\mathcal{A}$.

We say that a closed subspace M of H *reduces* an operator $S \in L(H)$ or S is *reduced* by M if both M and $M^\perp$ are invariant under S. We have the following equivalences.

Proposition 4.4. *Let* $\mathbf{S} \in L(H)$ *and* M *be a closed subspace of* H. *The following are equivalent.*

   (i) M *reduces* $\mathbf{S}$,
   (ii) $M^\perp$ *reduces* $\mathbf{S}$,
   (iii) M *reduces* $\mathbf{S}^*$,
   (iv) M *is invariant under* $\mathbf{S}$ *and* $\mathbf{S}^*$.

Proof.   Note $\mathbf{S}(M) \subset M$ iff $\mathbf{S}^*(M^\perp) \subset M^\perp$ and we have $M^{\perp\perp} = M$. $\square$

If $\mathbf{S} \in L(H)$ and M is a closed subspace of H, we denote the restriction of $\mathbf{S}$ to M by $\mathbf{S}|_M$.

Proposition 4.5. *If* M *reduces* $\mathbf{S}$, *then* $(\mathbf{S}|_M)^* = \mathbf{S}^*|_M$.

Proof. Set $\mathbf{U} = \mathbf{S}\,|_M$ and $\mathbf{V} = \mathbf{S}^*\,|_M$. Then $\mathbf{U}, \mathbf{V} \in L(M)$ by Proposition 4.4. For $x, y \in M$,

$$\mathbf{U}^* x \cdot y = x \cdot \mathbf{U}y = x \cdot \mathbf{S}y = \mathbf{S}^* x \cdot y = \mathbf{V}x \cdot y$$

so $\mathbf{U}^* = \mathbf{V}$. □

Corollary 4.6. *If M reduces S and S is normal, then $S\,|_M$ is normal.*

Proof. By Proposition 4.5,

$$(\mathbf{S}^*\mathbf{S})\,|_M = (\mathbf{S}^*\,|_M)(\mathbf{S}\,|_M) = (\mathbf{S}\,|_M^*)(\mathbf{S}\,|_M)$$
$$= (\mathbf{S}\mathbf{S}^*\,|_M) = (\mathbf{S}\,|_M)(\mathbf{S}^*\,|_M) = (\mathbf{S}\,|_M)(\mathbf{S}\,|_M)^*. \qquad \square$$

Lemma 4.7. *Suppose $0 \in \sigma(\mathbf{N})$. Given $\epsilon > 0$ there exists a closed subspace $M \neq \{0\}$ such that any operator which commutes with $\mathbf{N}\mathbf{N}^*$ is reduced by M and $\|\mathbf{N}\,|_M\| \leq \epsilon$.*

Proof. Set $\mathbf{A} = \mathbf{N}\mathbf{N}^*$. Since $0 \in \sigma(\mathbf{N})$, there exist $x_k \in H, \|x_k\| = 1$, such that $\|\mathbf{N}x_k\| \to 0$. Thus, $\mathbf{A}x_k \to 0$ and the self adjoint operator $\mathbf{A}$ has $0 \in \sigma(\mathbf{A})$. Let $\epsilon > 0$ and define the continuous function $f : \mathbb{R} \to \mathbb{R}$ by $f(t) = 1$ for $|t| \leq \epsilon/2$, $f(t) = 0$ for $|t| \geq \epsilon$ and $f(t) = 2(1 - |t/2|)$ for $\epsilon/2 < |t| < \epsilon$. Use the functional calculus of Theorem 2.2 to define $f(\mathbf{A})$. Let $M = \{x \in H : f(\mathbf{A})x = x\}$ so M is a closed subspace of H. Recall that if $\mathbf{B} \in L(H)$ commutes with $\mathbf{A}$, then $\mathbf{B}$ commutes with $f(\mathbf{A})$ (Theorem 2.2). Therefore, for $x \in M$,

$$\mathbf{B}x = \mathbf{B}f(\mathbf{A})x = f(\mathbf{A})\mathbf{B}x$$

so M is invariant under $\mathbf{B}$. Since $\mathbf{B}^*$ also commutes with $\mathbf{A}$, M is invariant under $\mathbf{B}^*$ and by Proposition 4.4 M reduces $\mathbf{B}$.

Since the functional calculus of $\mathbf{A}$ is an isometry, if $x \in M$ and $\|x\| = 1$, then

$$\|\mathbf{A}x\| = \|\mathbf{A}f(\mathbf{A})x\| \leq \|\mathbf{A}f(\mathbf{A})\| = \sup\{|tf(t)| : t \in \sigma(\mathbf{A})\} \leq \epsilon.$$

Thus, if $\|x\| = 1$,

$$\|\mathbf{N}x\|^2 = \mathbf{A}x \cdot x \leq \|\mathbf{A}x\| \leq \epsilon$$

so $\|\mathbf{N}\,|_M\| \leq \sqrt{\epsilon}$.

It remains to show that $M \neq \{0\}$. Now

$$\|(I - f(\mathbf{A}))f(2\mathbf{A})\| = \sup\{|1 - f(t)|\,|f(2t)| : t \in \sigma(\mathbf{A})\} = 0$$

since $f(t) = 1$ when $f(2t) \neq 0$. Therefore, every element in the range of the operator $f(2\mathbf{A})$ lies in M, and this range is not $\{0\}$ since

$$\|f(2\mathbf{A})\| = \sup\{|f(2t)| : t \in \sigma(\mathbf{A})\} \geq |f(0)| = 1. \qquad \square$$

We next establish a version of the spectral mapping theorem for the functional calculus Ψ.

Theorem 4.8. *Let $p(s, t)$ be a complex polynomial in 2 variables. Then*

$$\sigma(p(\mathbf{N}, \mathbf{N}^*)) = \{p(z, \overline{z}) : z \in \sigma(\mathbf{N})\}.$$

Proof. Let $p(s, t) = \sum a_{ij} s^i t^j$. Let $\lambda \in \sigma(\mathbf{N})$. Then there exist $x_k \in H, \|x_k\| = 1$, such that $\|(\lambda - \mathbf{N})x_k\| \to 0$. Then $\|(\overline{\lambda} - \mathbf{N}^*)x_k\| \to 0$. Therefore,

$$(p(\mathbf{N}, \mathbf{N}^*) - p(\lambda, \overline{\lambda}))x_k = \sum a_{ij}(\mathbf{N}^i \mathbf{N}^{*j} - \lambda^i \overline{\lambda}^j)x_k$$

$$= \sum a_{ij}(\mathbf{N}^i(\mathbf{N}^{*j} - \overline{\lambda}^j)x_k + \overline{\lambda}^j(\mathbf{N}^i - \lambda^i)x_k)$$

$$= \sum a_{ij}(\mathbf{N}^i(\mathbf{N}^{*(j-1)} + \dots + \overline{\lambda}^{j-1})(\mathbf{N}^* - \overline{\lambda})$$

$$+ \overline{\lambda}^j(\mathbf{N}^{i-1} + \dots + \lambda^{i-1})(\mathbf{N} - \lambda))x_k \to 0.$$

Thus, $p(\lambda, \overline{\lambda}) \in \sigma(p(\mathbf{N}, \mathbf{N}^*))$ and $\sigma(p(\mathbf{N}, \mathbf{N}^*)) \supset \{p(z, \overline{z}) : z \in \sigma(\mathbf{N})\}$.

Suppose $\mu \in \sigma(p(\mathbf{N}, \mathbf{N}^*))$. The operator $\mathbf{B} = p(\mathbf{N}, \mathbf{N}^*) - \mu$ is normal and has $0 \in \sigma(\mathbf{B})$. by Lemma 4.7 for each n there is a closed subspace $M_n \neq \{0\}$ which reduces $\mathbf{B}$ and has $\|\mathbf{B}\|_{M_n}\| \leq 1/n$ and since $\mathbf{N}$ commutes with $\mathbf{B}$, each M_n also reduces $\mathbf{N}$. Therefore, $\mathbf{N}|_{M_n}$ is normal by Corollary 4.6. Choose $\lambda_n \in \sigma(\mathbf{N}_{M_n})$. Then there exists $y_n \in M_n, \|y_n\| = 1$, such that $\|(\lambda_n - \mathbf{N})y_n\| \leq 1/n$. The sequence $\{\lambda_n\}$ is bounded by $\|\mathbf{N}\|$ and, therefore, has a convergent subsequence, still denoted by $\{\lambda_n\}$, which converges to some $\lambda \in \mathbb{C}$. Now, $\lambda \in \sigma(\mathbf{N})$ since

$$\|(\lambda - \mathbf{N})y_n\| \leq |\lambda_n - \lambda| + \|(\lambda_n - \mathbf{N})y_n\| \to 0.$$

But, in the first part we showed that if $(\lambda - \mathbf{N})y_n \to 0$, then

$$(p(\mathbf{N}, \mathbf{N}^*) - p(\lambda, \overline{\lambda}))y_n \to 0.$$

Now $y_n \in M_n$ so

$$\|By_n\| = \|(p(\mathbf{N}, \mathbf{N}^*) - \mu)y_n\| \leq 1/n$$

and $\mu = p(\lambda, \overline{\lambda})$. Thus, $\sigma(p(\mathbf{N}, \mathbf{N}^*)) \subset \{p(z, \overline{z}) : z \in \sigma(\mathbf{N})\}$. $\square$

From Theorem 4.8 we can obtain that the functional calculus Ψ is an isometry.

Corollary 4.9. *If $\mathbf{N} \in L(H)$ is normal, then*

$$\|p(\mathbf{N}, \mathbf{N}^*)\| = \sup\{|p(z, \overline{z})| : z \in \sigma(\mathbf{N})\}.$$

Thus, if $\Psi : \mathcal{P} \to \mathcal{A}$ is the functional calculus defined by

$$\Psi p = p(\mathbf{N}, \mathbf{N}^*),$$

then Ψ is an isometric, algebra isomorphism from $\mathcal{P}$ into $\mathcal{A}$, and since $\mathcal{P}$ is dense in $C_{\mathbb{C}}(\sigma(\mathbf{N}))$, Ψ can be extended to an isometric, algebra isomorphism of $C_{\mathbb{C}}(\sigma(\mathbf{N}))$ onto $\mathcal{A}$. Moreover, $\Psi(\overline{f}) = \Psi(f)^*$ for every $f \in C_{\mathbb{C}}(\sigma(\mathbf{N}))$ and if $S \in L(H)$ commutes with $\mathbf{N}$ and $\mathbf{N}^*$, then S commutes with Ψf for every $f \in C_{\mathbb{C}}(\sigma(\mathbf{N}))$.

We can now repeat the proof of the spectral theorem for self adjoint operators given in Theorem 2.7 to obtain the spectral theorem for normal operators as given in Theorem 2.7.

We now give several applications of the functional calculus for normal operators. First, we show the resolution of the identity can be used to characterize eigenvalues.

Theorem 4.10. *Let $\mathbf{N}$ be normal with resolution of the identity E. Then $\mu \in \sigma(\mathbf{N})$ is an eigenvalue iff $E\{\mu\} \neq 0$.*

Proof. Suppose $E\{\mu\} \neq 0$ and pick $x \neq 0$ such that $E\{\mu\}x = x$. Then

$$\mathbf{N}x = \int_{\sigma(\mathbf{N})} z dE(z)E\{\mu\}x = \int_{\sigma(\mathbf{N})} z dE(z \cap \{\mu\})x = \mu x$$

so μ is an eigenvalue with eigenvector x. Suppose $\mathbf{N}x = \mu x$ with $x \neq 0$. Define bounded Baire functions f_n by $f_n(z) = 1/(\mu - z)$ if $|\mu - z| > 1/n$ and $f_n(z) = 0$ if $|\mu - z| \leq 1/n$. Then $f_n(z)(\mu - z) = \chi_{A_n}(z)$, where $A_n = \{z : |\mu - z| \geq 1/n\}$ so $f_n(\mathbf{N})(\mu - \mathbf{N}) = E(A_n)$. Using the countable additivity of $E(\cdot)x$ and letting $n \to \infty$ gives $E\{z : z \neq \mu\}x = 0$ so $E\{\mu\}x = x \neq 0$. $\square$

Remark 4.11. Note we have shown that $\ker(\mu - \mathbf{N}) = rangeE\{\mu\}$.

Unitary Operators.

Using the functional calculus for self adjoint and normal operators, we can obtain an interesting characterization of unitary operators.

Theorem 4.12. *An operator $U \in L(H)$ is unitary iff*

$$U = e^{iA}$$

for some self adjoint operator $\mathbf{A} \in L(H)$.

Proof.   Suppose $\mathbf{A}$ is self adjoint and let E be the resolution of the identity for $\mathbf{A}$ with $\sigma = \sigma(\mathbf{A})$. Then

$$e^{i\mathbf{A}} = \int_\sigma e^{it} dE(t)$$

so from Theorem 1.13,

$$(e^{i\mathbf{A}})(e^{i\mathbf{A}})^* = \int_\sigma e^{it} dE(t) \int_\sigma e^{-it} dE(t) = \int_\sigma dE = I$$

and $e^{i\mathbf{A}}$ is unitary.

For the converse, since $\mathbf{U}$ is normal, $\mathbf{U}$ has a resolution of the identity G as in Theorem 4.3. Since $\sigma(\mathbf{U}) \subset \{z : |z| = 1\}$, there exists a real valued, bounded Baire function $f : \sigma(\mathbf{U}) \to \mathbb{R}$ such that $\exp(if(z)) = z$ for every $z \in \sigma(\mathbf{U})$. Put $\mathbf{A} = f(\mathbf{U})$ and note $\mathbf{A}$ is self adjoint since f is real valued. Then

$$e^{i\mathbf{A}} = e^{if(\mathbf{U})} = \mathbf{U}$$

by properties of the functional calculus (Theorem 1.15).   □

In Theorem C.28 we showed that the spectrum of a unitary operator is always contained in $\{z \in \mathbb{C} : |z| = 1\}$. We can use the functional calculus to establish the converse of this statement for normal operators.

Theorem 4.13. *A normal operator* $\mathbf{T}$ *is unitary iff*

$$\sigma(\mathbf{T}) \subset \{z \in \mathbb{C} : |z| = 1\}.$$

Proof.   Suppose $\sigma(\mathbf{T}) \subset \{z \in \mathbb{C} : |z| = 1\}$. Let E be the resolution of the identity for $\mathbf{T}$. For $x, y \in H$,

$$\mathbf{T}^*\mathbf{T}x \cdot y = \mathbf{T}\mathbf{T}^*x \cdot y = \int_{\sigma(\mathbf{T})} z\bar{z} d(E(z)x \cdot y) = \int_{\sigma(\mathbf{T})} d(E(z)x \cdot y) = Ix \cdot y$$

which implies $\mathbf{T}^*\mathbf{T} = \mathbf{T}\mathbf{T}^* = I$. Hence, $\mathbf{T}$ is unitary.

The converse was established in Theorem C.28.   □

Of course, since every unitary operator is normal, a unitary operator has a spectral representation as in Theorem 4.3. However, the spectral theorem for unitary operators is usually given in a different form which we now describe.

Theorem 4.14. *Let* $\mathbf{U} \in L(H)$ *be unitary. There exists a unique spectral measure* F *on* $\mathcal{B}(0, 2\pi]$ *such that*

$$\mathbf{U} = \int_0^{2\pi} e^{it} dF(t).$$

Proof. Let G be the resolution of the identity for **U** as given in Theorem 4.3. Let $D = \{z : |z| = 1\}$ and $J = (0, 2\pi]$. Extend G from $\mathcal{B}(\sigma(\mathbf{U}))$ to $\mathcal{B}(D)$ by setting $\widehat{G}(A) = G(A \cap \sigma(\mathbf{U}))$ for $A \in \mathcal{B}(D)$. If $f \in \mathcal{B}(D)$, then

$$\int_D f d\widehat{G} = \int_{\sigma(\mathbf{U})} f dG.$$

Define a homeomorphism h from J to D by $h(t) = e^{it}$ and set $F = \widehat{G} \circ h^{-1}$. From Theorem 1.16, if $f : D \to \mathbb{C}$ is a bounded, Baire function, then

$$\int_D f d\widehat{G} = \int_0^{2\pi} f(e^{it}) dF(t).$$

In particular,

$$\mathbf{U} = \int_D z d\widehat{G} = \int_0^{2\pi} e^{it} dF(t). \qquad \square$$

Every complex number has a polar decomposition $z = re^{i\theta}$. We consider an analogous decomposition for operators. That is, we seek a decomposition in the form $\mathbf{T} = \mathbf{PU}$ where $\mathbf{P}$ is positive and $\mathbf{U}$ is unitary.

Theorem 4.15. *(Polar Decomposition) If* $\mathbf{N} \in L(H)$ *is normal, then there exist a positive operator* $\mathbf{P}$ *and a unitary operator* $\mathbf{U}$ *such that* $\mathbf{N} = \mathbf{PU}$. *Moreover,* $\mathbf{P}$ *and* $\mathbf{U}$ *commute and both commute with* $\mathbf{N}$.

Proof. Define bounded, Baire functions $p(z) = |z|$ and $u(z) = z/|z|$ if $z \neq 0$ and $u(0) = 1$. Set $\mathbf{P} = p(\mathbf{N})$ and $\mathbf{U} = u(\mathbf{N})$. Since $p \geq 0$, $\mathbf{P}$ is positive, and $u\overline{u} = \overline{u}u = 1$ implies $\mathbf{UU}^* = \mathbf{U}^*\mathbf{U} = I$ so $\mathbf{U}$ is unitary. Since $(pu)(z) = z$, $\mathbf{PU} = \mathbf{N}$. The commutativity statement follows from the properties of the functional calculus. $\qquad \square$

There is a similar polar decomposition for invertible operators

Theorem 4.16. *If* $\mathbf{T} \in L(H)$ *is invertible, then there exist a positive operator* $\mathbf{P}$ *and a unitary operator* $\mathbf{U}$ *such that* $\mathbf{T} = \mathbf{UP}$. *The decomposition is unique.*

Proof. Since $\mathbf{T}$ is invertible so are $\mathbf{T}^*$ and $\mathbf{T}^*\mathbf{T}$ so the positive square root $\mathbf{P}$ of $\mathbf{T}^*\mathbf{T}$ is also invertible. Set $\mathbf{U} = \mathbf{TP}^{-1}$. Then $\mathbf{U}$ is invertible and

$$\mathbf{U}^*\mathbf{U} = \mathbf{P}^{-1}\mathbf{T}^*\mathbf{TP}^{-1} = \mathbf{P}^{-1}\mathbf{P}^2\mathbf{P}^{-1} = I$$

so $\mathbf{U}$ is unitary. Since $\mathbf{P}$ is invertible, $\mathbf{U}$ must be $\mathbf{TP}^{-1}$ and $\mathbf{P} = \sqrt{\mathbf{T}^*\mathbf{T}}$. $\qquad \square$

The functional calculus and the polar decomposition can also be used to give information about the group of invertible operators on a Hilbert space.

Theorem 4.17. *Let $\mathcal{G}$ be the group of invertible elements in $L(H)$. Then $\mathcal{G}$ is path connected and every $T \in \mathcal{G}$ is the product of two exponentials.*

Proof. Let $\mathbf{T} \in \mathcal{G}$ and $\mathbf{T} = \mathbf{UP}$ be the polar decomposition. Since $\mathbf{P}$ is positive and invertible, $\sigma(\mathbf{P}) \subset (0, \infty)$, ln is a continuous real function on $\sigma(\mathbf{P})$. From the functional calculus there is a self adjoint operator $\mathbf{A}$ such that $\mathbf{P} = e^{\mathbf{A}}$. Since $\mathbf{U}$ is unitary, $\sigma(\mathbf{U}) \subset \{z \in \mathbb{C} : |z| = 1\}$ so there is a bounded, Baire function f on $\sigma(\mathbf{U})$ such that $\exp(if(z)) = z$ for $z \in \sigma(\mathbf{U})$. Set $\mathbf{Q} = f(\mathbf{U})$. Then $\mathbf{Q}$ is self adjoint and $\mathbf{U} = \exp(i\mathbf{Q})$ so

$$\mathbf{T} = \mathbf{UP} = \exp(i\mathbf{Q})\exp(\mathbf{A}).$$

Now define

$$\mathbf{T}_r = \exp(ir\mathbf{Q})\exp(\mathbf{A})$$

for $0 \leq r \leq 1$ so $r \to \mathbf{T}_r$ is a continuous map from $[0,1]$ into $\mathcal{G}$ such that $\mathbf{T}_0 = I$ and $\mathbf{T}_1 = \mathbf{T}$. Hence, $\mathcal{G}$ is path connected. $\qquad\square$

Finally, we can use the spectral theorem for unitary operators to prove an abstract ergodic theorem.

Theorem 4.18. *If $\mathbf{U}$ is unitary and $x \in H$, then the averages*

$$\mathbf{A}_n x = \sum_{j=0}^{n-1} \mathbf{U}^j x/n$$

converge in norm to some $y \in H$.

Proof. Let E be the resolution of the identity for $\mathbf{U}$. Define a_n and b on $\{z \in \mathbb{C} : |z| = 1\}$ by

$$a_n(z) = \sum_{j=0}^{n-1} z^j/n$$

and $b(z) = 0$ for $z \neq 1$, $b(1) = 1$. Then $\mathbf{A}_n = a_n(\mathbf{U})$. Set $y = b(\mathbf{U})x$ so

$$\|y - \mathbf{A}_n x\|^2 = \|b(\mathbf{U})x - a_n(\mathbf{U})\|^2 = \int_{\sigma(\mathbf{U})} |b(z) - a_n(z)|^2 \, dEx \cdot x.$$

Since $|b - a_n| < 1$ and $(b - a_n)(z) \to 0$ on $\{z \in \mathbb{C} : |z| = 1\}$, the Dominated Convergence Theorem implies

$$\|y - \mathbf{A}_n x\| \to 0.$$
$\qquad\square$

For a discussion of ergodic theorems, see [DS], VIII.4-5 and [RN], sections 144-145.

Remark 4.19. The proof of the spectral theorem given in Theorem 4.3 is like that given in [Be2] based on the product of spectral measures. There are many other proofs. For example, see [RN] for a spectral representation as a Riemann-Stieltjes type integral. Many proofs are based on the Gelfand map for commutative Banach algebras; see [TL]. Likewise, there are other proofs of the spectral theorem for unitary operators; see [RN], [TL].

Chapter 5

Integrating Vector Valued Functions

In this chapter we establish the results on the integration of vector valued functions with respect to positive measures which will be needed in subsequent chapters. If one is interested in generalizing properties/operations of scalar valued functions to functions with values in a normed linear space (or locally convex spaces), there are two natural procedures which can be followed-often referred to as "weak" or "strong". In the weak procedure, if f is a function with values in a normed space X and one is interested in seeking to extend a scalar property/operation to f, it is simply required that the scalar function $x'f = x' \circ f$ possesses the scalar property for every $x' \in X'$ and scalar operations are performed on $x'f$. In the strong procedure, one seeks to imitate the scalar definition/operation to f by replacing absolute values in the scalar field with norms, $\|\cdot\|$, in the normed space. In this chapter we will carry out both of these procedures for the scalar notions of measurability and integration. It will be seen that these two procedures produce very different theories in this case. See, however, the case for analytic functions considered in Chapter 7.

5.1 Vector Valued Measurable Functions

In this section we will apply the "weak" and "strong" extension principles to define measurability for vector valued functions.

Let X be a Banach space, Σ a σ-algebra of subsets of a set S and $\mu : \Sigma \to [0, \infty]$ a σ-finite measure. A function $g : S \to X$ is Σ-simple (or just *simple* if Σ is understood) if g has a representation $g = \sum_{j=1}^{n} x_j \chi_{A_j}$, where $x_j \in X$ and $A_j \in \Sigma$. A function $f : S \to X$ is scalarly Σ-measurable (some authors prefer weakly measurable) if $x'f$ is Σ-measurable for every $x' \in X'$ (if Σ is understood, we just say f is *scalarly measurable*). A simple function is obviously scalarly measurable. A function $f : S \to X$ is strongly Σ-measurable (or just *strongly measurable*) if there exists a sequence of Σ-simple functions $\{g_j\}$ such that $g_j \to f$ (in norm) $\mu - a.e.$ A strongly measurable function is obviously scalarly measurable but Example 5.2 below shows the converse is false.

Proposition 5.1. *If $f : S \to X$ is strongly measurable, then the scalar function $\|f(\cdot)\|$ is measurable.*

Proof. If $g = \sum_{j=1}^n x_j \chi_{A_j}$ is a simple function with the $\{A_j\}$ pairwise disjoint, then $\|g(\cdot)\| = \sum_{j=1}^n \|x_j\| \chi_{A_j}$ is measurable. Therefore, if $\{g_j\}$ is a sequence of simple functions converging $\mu - a.e.$ to f, then $\|g_j(\cdot)\| \to \|f(\cdot)\|$ $\mu - a.e.$ so $\|f(\cdot)\|$ is measurable. $\qquad\qquad\square$

Example 5.2. Let $S = [0,1]$ and let λ be counting measure on S. Put $H = L^2(\lambda) = l^2(S)$. Then H is a non-separable Hilbert space and $\{e_t : t \in S\}$, $e_t = \chi_{\{t\}}$ is a complete orthonormal subset of H. Let μ be Lebesgue measure on S and let P be a non-measurable subset of S. Define $f : S \to H$ by $f(t) = e_t$ if $t \in P$ and $f(t) = 0$ otherwise. Now f is scalarly measurable since if $h \in H' = H$, then h has a Fourier expansion $h = \sum_{t \in S}(h \cdot e_t)e_t$ with $\{t : h \cdot e_t \neq 0\}$ countable and $h \cdot f(t) = 0$ if $t \notin P$ and $h \cdot f(t) = h \cdot e_t$ if $t \in P$ so $h \cdot f(\cdot)$ is 0 except at countably many points and is measurable. However, $\|f(\cdot)\| = \chi_P$ is not measurable so f is not strongly measurable by Proposition 5.1.

Note that in this example f has non-separable range. We next establish an important result of Pettis relating scalar and strong measurability.

Definition 5.3. A subset $\Gamma \subset X'$ is a norming set for X if

$$\|x\| = \sup\{|x'(x)| : x' \in \Gamma\}$$

for every $x \in X$.

For example, the unit ball $\{x' \in X' : \|x'\| \leq 1\}$ is always a norming set for X. However, proper subsets of the unit ball are often norming subsets. For example, if S is a compact, Hausdorff space and δ_t is the Dirac (point) measure at t (Example E.3), then $\{\delta_t : t \in D\}$ is a norming subset for $C(S)$ for any dense subset D of S. For separable spaces we have:

Theorem 5.4. *If Y is a separable normed space, then Y' contains a countable norming set for Y.*

Proof. Let $\{y_j\}$ be a countable dense subset of Y. For each j there exists $y'_j \in Y'$, $\|y'_j\| = 1$, such that $|y'_j(y_j)| = \|y_j\|$. Put $\Gamma = \{y'_j : j \in \mathbb{N}\}$. If $y \in Y$ and $\epsilon > 0$, there exists j such that

$$|\|y\| - \|y_j\|| \le \|y - y_j\| < \epsilon.$$

Then

$$\|y\| \ge |y'_j(y)| \ge |y'_j(y_j)| - |y'_j(y - y_j)| \ge \|y_j\| - \epsilon \ge \|y\| - 2\epsilon.$$

Hence,

$$\|y\| = \sup\{|y'_j(y)| : j \in \mathbb{N}\}$$

and Γ is a norming subset for Y. $\qquad\square$

Proposition 5.5. *Suppose $\Gamma \subset X'$ is a norming subset for X. If $f : S \to X$ is such that $x'f$ is measurable for every $x' \in \Gamma$, the $\|f(\cdot)\|$ is measurable.*

Proof. For each $t \in S$,

$$\|f(t)\| = \sup\{|x'(f(t))| : x' \in \Gamma\}$$

so the result is immediate. $\qquad\square$

A function $g : S \to X$ is *countably valued* if g has the form $g = \sum_{j=1}^{\infty} x_j \chi_{A_j}$, where $x_j \in X$ and $\{A_j\} \subset \Sigma$ are pairwise disjoint. Such a function is clearly strongly measurable.

Proposition 5.6. *Let X be separable and $f : S \to X$ be scalarly measurable. Then for every $\epsilon > 0$ there exists a countably valued function $g : S \to X$ such that $\|g(t) - f(t)\| < \epsilon$ for every $t \in S$.*

Proof. Since X is separable, X can be covered by a countable number of open spheres

$$S_n = S(x_n, \epsilon) = \{x : \|x - x_n\| < \epsilon\}.$$

By Proposition 5.5 and Theorem 5.4 the function $t \to \|f(t) - x_n\|$ is measurable so

$$B_n = \{t \in S : f(t) \in S_n\} \in \Sigma$$

and $S = \cup_{n=1}^{\infty} B_n$. For each n disjointify the $\{B_n\}$ by setting $A_1 = B_1$ and

$$A_n = B_n \setminus \bigcup_{j=1}^{n-1} B_j$$

for $n \ge 2$. Define $g : S \to X$ by

$$g = \sum_{n=1}^{\infty} x_n \chi_{A_n}.$$

Then g is countably valued and $\|g(t) - f(t)\| < \epsilon$ for every $t \in S$. $\qquad\square$

Proposition 5.7. *Let $f_n : S \to X$ be strongly measurable for every $n \in \mathbb{N}$ and $f : S \to X$. If $f_n \to f$ $\mu - a.e.$, then f is strongly measurable.*

Proof. Pick $h \in L^1(\mu)$ such that $h(t) > 0$ for every $t \in S$. Fix m. Now

$$\|f_n(t) - f_m(t)\| \to \|f(t) - f_m(t)\|$$

for μ almost all $t \in S$, so $\|f(\cdot) - f_m(\cdot)\|$ is measurable (Proposition 5.1). Hence,

$$\epsilon_m = \int_S h(t) \|f(t) - f_m(t)\| / (1 + \|f(t) - f_m(t)\|) d\mu(t) \to 0$$

by the Dominated Convergence Theorem. For each m there exists a sequence of simple functions $\{g_{mj}\}$ such that

$$\lim_j g_{mj} = f_m \ \mu - a.e.$$

By the argument above there exists a simple function g_m such that

$$\int_S h(t) \|g_m(t) - f_m(t)\| / (1 + \|g_m(t) - f_m(t)\|) d\mu(t) < 1/m.$$

Hence,

$$\int_S h(t) \|f(t) - g_m(t)\| / (1 + \|f(t) - g_m(t)\|) d\mu(t) \leq \epsilon_m + 1/m.$$

By Riesz's Theorem ([Sw1] 3.6.6), there is a subsequence $\{g_{m_k}\}$ converging to f $\mu - a.e.$ $\qquad\square$

From Propositions 5.6 and 5.7 we see that a function $f : S \to X$ is strongly measurable iff f is the uniform limit of a sequence of countably valued functions.

A function $f : S \to X$ is *μ-almost separably valued* if there exists a μ null set $A \in \Sigma$ such that $f(S \backslash A)$ is separable. We now have Pettis' theorem.

Theorem 5.8. *(Pettis) Let $f : S \to X$. Then f is strongly measurable iff f is scalarly measurable and μ-almost separably valued.*

Proof. $\Rightarrow$: There exist a μ null set A and a sequence of simple functions $\{g_j\}$ such that $g_j(t) \to f(t)$ for $t \in S \backslash A$. Since $f(S \backslash A)$ is contained in the closure of $\cup_{j=1}^{\infty} range(g_j)$, f is μ-almost separably valued. As observed earlier, f is scalarly measurable.

$\Leftarrow$: Without loss of generality, we may assume X is separable. The result now follows from Propositions 5.6 and 5.7. $\qquad\square$

For a result more analogous to the "usual" scalar definition of measurablity, we have:

Theorem 5.9. *Let $f : S \to X$. Then f is strongly measurable iff*

    (i) f is μ-almost separably valued and
    (ii) $f^{-1}(G) \in \Sigma$ for every open set $G \subset X$.

Proof. $\Rightarrow$: (i) holds by Theorem 5.8 so we may assume X is separable. Since $\|f(\cdot) - x\|$ is measurable for every $x \in X$,

$$f^{-1}(S(x, r)) = \{t : \|f(t) - x\| < r\} \in \Sigma$$

for every $r > 0$. Since X is separable, every open set G in X is a countable union of open spheres so $f^{-1}(G) \in \Sigma$ by the observation above. Hence, (ii) holds.

$\Leftarrow$: Condition (ii) implies that $x'f$ is measurable for every $x' \in X'$ since

$$(x'f)^{-1}(G) = f^{-1}((x')^{-1}(G)) \in \Sigma$$

by the continuity of x'. Therefore, f is strongly measurable by Theorem 5.8. $\qquad\qquad\square$

5.2   Integrating Vector Valued Functions

Using the "weak" extension principle discussed earlier in this chapter, we say that a function $f : S \to X$ is *scalarly μ-integrable* (weakly μ-integrable) if $x'f$ is μ-integrable for every $x' \in X'$. The following theorem is used in defining the Dunford integral.

Theorem 5.10. *Let $f : S \to X$ be scalarly μ-integrable. The map $F : X' \to L^1(\mu)$ defined by $F(x') = \int_S x'f d\mu$ is linear and continuous.*

Proof. First assume μ is finite and let $A_k = \{t \in S : \|f(t)\| \leq k\}$ so $A_k \uparrow S$. Set $f_k = \chi_{A_k} f$ and define $F_k : X' \to L^1(\mu)$ by

$$F_k(x') = \int_S x'f_k d\mu.$$

F_k is obviously linear and is continuous since

$$\|F_k(x')\|_1 = \int_S |x'f_k| \, d\mu \leq \|x'\| \, k\mu(A_k).$$

Since $x'f_k \to x'f$ pointwise and $|x'f_k| \le |x'f|$, the Dominated Convergence Theorem implies

$$\|F_k(x') - F(x')\|_1 \to 0.$$

Hence, F is linear and continuous by the Uniform Boundedness Principle.

If μ is σ-finite, let $B_k \in \Sigma$ be chosen such that $B_k \uparrow S$ with $\mu(B_k) < \infty$. Set $g_k = \chi_{B_k} f$ and define $G_k : X' \to L^1(\mu)$ by

$$G_k(x') = \int_S x' g_k d\mu.$$

By the paragraph above, G_k is linear and continuous and as above $\|G_k(x') - F(x')\|_1 \to 0$ so F is linear and continuous by the Uniform Boundedness Principle. □

Alternately, one can show that the map F has a closed graph and is continuous by the Closed Graph Theorem.

The transpose $F' : L^\infty(\mu) \to X''$ is given by

$$F'(g)(x') = g(F(x')) = \int_S gx'f d\mu.$$

We then define the integral of f with respect to μ over $A \in \Sigma$ to be the element $\int_A f d\mu = F'(\chi_A) \in X''$ so

$$\int_A f d\mu(x') = \int_A x'f d\mu \ for \ x' \in X'.$$

The element $\int_A f d\mu \in X''$ is called the *Dunford integral* (sometimes the Gelfand integral) of f. Of course, this integral has the unpleasant feature that although f has values in X the integral has values in the bidual X''. We single out the functions whose integrals have values in X; we say that f is *Pettis integrable* if f is Dunford integrable and $\int_A f d\mu \in X$ for every $A \in \Sigma$. If $g = \sum_{k=1}^n x_k \chi_{A_k}$ is a Σ-simple function, then g is Pettis integrable with respect to μ iff $\mu(A_k) < \infty$ when $x_k \ne 0$ and in this case $\int_A f d\mu = \sum_{k=1}^n \mu(A \cap A_k) x_k$. Of course, when X is reflexive every Dunford integrable function is Pettis integrable, but, in general, this is not the case.

Example 5.11. Let μ be counting measure on $\mathbb{N}$. define $f : \mathbb{N} \to c_0$ by $f(k) = e^k$. Let $x' = \{t_k\} \in l^1 = (c_0)'$. Then $x'f(k) = t_k$ so f is scalarly μ-integrable with

$$\int_A x'f d\mu = \sum_{k \in A} t_k = x'(\chi_A).$$

Hence, $\int_A f d\mu = \chi_A \in l^\infty$ and when A is infinite, $\chi_A \notin c_0$ so f is not Pettis integrable.

From continuity properties of the transpose map we can obtain additivity properties for the Dunford and Pettis integrals.

Theorem 5.12. *Let $f : S \to X$ be scalarly μ-integrable. The indefinite integral*

$$\int_{.} f d\mu : A \to \int_A f d\mu$$

from Σ to X'' is weak* countably additive. If f is Pettis integrable, then the indefinite integral $\int_{.} f d\mu : A \to \int_A f d\mu$ from Σ to X is norm countably additive.

Proof. If $\{A_j\}$ is a pairwise disjoint sequence from Σ with union A, then the series $\sum_{j=1}^{\infty} \chi_{A_j}$ is weak* convergent in $L^{\infty}(\mu)$ to χ_A. Since F' is weak*-weak* continuous, the series $\sum_{j=1}^{\infty} \int_{A_j} f d\mu$ is weak* convergent in X'' to $\int_A f d\mu$.

The last statement follows from the Orlicz-Pettis Theorem (A.6) □

Note the indefinite integral in Example 5.11 is not norm countably additive.

Let $\mathcal{D}(\mu, X)$ $[\mathcal{P}(\mu, X)]$ be the space of all X valued functions on S which are Dunford [Pettis] μ-integrable. Then $\mathcal{D}(\mu, X)$ $[\mathcal{P}(\mu, X)]$ is a vector space with the usual operations of pointwise addition and scalar multiplication. The space $\mathcal{D}(\mu, X)$ has a natural semi-norm defined by

$$\|f\|_P = \sup\{ \int_S |x' f| \, d\mu : \|x'\| \leq 1 \};$$

note $\|f\|_P = \|F\| < \infty$ by Proposition 5.1. In general, the space $\mathcal{D}(\mu, X)$ $[\mathcal{P}(\mu, X)]$ is not complete under this semi-norm.

Example 5.13. Let μ be Lebesgue measure on $(0, 1]$ and $X = L^2(\mu)$. We show $\mathcal{D}(\mu, X) = \mathcal{P}(\mu, X)$ is not complete under $\|\cdot\|_p$. Let $\{x_{ij} : 1 \leq j \leq 2^i, i \in \mathbb{N}\}$ be a complete orthonormal set in X and set

$$E_{ij} = ((j-1)/2^i, j/2^i], 1 \leq j \leq 2^i, i \in \mathbb{N}.$$

Set

$$y_i = \sum_{j=1}^{2^i} x_{ij} \chi_{E_{ij}}.$$

For $x' \in X'$,

$$\int_0^1 |x'y_i| \, d\mu = \sum_{j=1}^{2^i} |x' \cdot x_{ij}|/2^i \leq (\sum_{j=1}^{2^i} |x' \cdot x_{ij}|^2)^{1/2} (\sum_{j=1}^{2^i} 1/2^{2i})^{1/2}$$

$$\leq \|x'\| \, 2^{-i/2}$$

by the Cauchy-Schwarz and Bessel inequalities. Hence, $\|y_i\|_P \leq 2^{-i/2}$. If $s_n = \sum_{i=1}^n y_i$, then

$$\|s_n - s_m\|_P \leq \sum_{i=m+1}^n 2^{-i/2} \qquad \text{for } n > m$$

so $\{s_n\}$ is a Cauchy sequence in $\mathcal{P}(\mu, X)$. Suppose there exists a (strongly measurable) function $f \in \mathcal{P}(\mu, X)$ such that $\|s_n - f\|_P \to 0$. Then for every $i, j, A \in \Sigma$,

$$x_{ij} \cdot \int_{A \cap E_{ij}} (s_n - f) d\mu \to 0 \text{ as } n \to \infty.$$

But,

$$x_{ij} \cdot \int_{A \cap E_{ij}} (s_n - f) d\mu = \int_{A \cap E_{ij}} x_{ij} \cdot s_n d\mu - \int_{A \cap E_{ij}} x_{ij} \cdot f d\mu$$

$$= \int_{A \cap E_{ij}} (1 - x_{ij} \cdot f) d\mu$$

so $x_{ij} \cdot f$ must equal 1 $\mu - a.e.$ in E_{ij}. This is impossible for then there would be a $t \in (0, 1]$ such that for infinitely many i, j, $x_{ij} \cdot f(t) = 1$, i.e., the Fourier coefficients of $f(t)$ would not converge to 0.

In this section we use the (strong) extension principle discussed earlier to define an integral, called the Bochner integral, for strongly measurable functions.

A Σ-simple function $g = \sum_{j=1}^n x_j \chi_{A_j} : S \to X$ is Bochner μ-integrable over $A \in \Sigma$ iff $\mu(A \cap A_j) < \infty$ for every $x_j \neq 0$ and in this case the Bochner integral of g over A is defined to be

$$\int_A g d\mu = \sum_{j=1}^n x_j \mu(A \cap A_j)$$

as in the case of the Pettis integral, so at this point our notation for the integral is unambiguous. When it is necessary to distinguish between various integrals, we will devise appropriate notation but throughout this section unless specified otherwise all integrals will be Bochner integrals. Note that a simple function g is Bochner μ-integrable iff $\|g(\cdot)\|$ is μ-integrable and, in this case,

$$\left\| \int_A g \, d\mu \right\| \leq \int_A \|g(\cdot)\| \, d\mu;$$

we will see later that this property carries forward to strongly measurable functions.

Definition 5.14. A strongly measurable function $f : S \to X$ is Bochner μ-integrable if

    (i) there exists a sequence of simple Bochner μ-integrable functions $\{g_j\}$ such that $g_j \to f$ $\mu - a.e.$ and

   (ii) $\lim \int_S \|f(\cdot) - g_j(\cdot)\| \, d\mu \to 0$.

The Bochner integral of f with respect to μ is defined to be

$$\int_S f \, d\mu = \lim \int_S g_j \, d\mu.$$

We say that f is *Bochner μ-integrable* over $A \in \Sigma$ if $\chi_A f$ is Bochner μ-integrable and define

$$\int_A f \, d\mu = \int_A \chi_A f \, d\mu.$$

A few comments are in order. First, note the integrals in (ii) make sense since the function $\|f(\cdot) - g_j(\cdot)\|$ is measurable (Proposition 5.1). Since

$$\left\| \int_S g_j \, d\mu - \int_S g_k \, d\mu \right\| \leq \int_S \|g_j(\cdot) - g_k(\cdot)\| \, d\mu$$

$$\leq \int_S \|f(\cdot) - g_j(\cdot)\| \, d\mu + \int_S \|f(\cdot) - g_k(\cdot)\| \, d\mu,$$

(ii) implies that $\{\int_S g_j \, d\mu\}$ is a Cauchy sequence in X so $\lim \int_S g_j \, d\mu$ exists. Moreover, this limit is independent of the sequence $\{g_j\}$ satisfying (i) and (ii) [if $\{h_j\}$ is another sequence satisfying (i) and (ii), consider the interlaced sequence $\{g_1, h_1, g_2, h_2, ...\}$ and note the limit of the integrals of the interlaced sequence must exist].

We have a very useful criterion for the existence of the Bochner integral.

Theorem 5.15. *Let $f : S \to X$ be strongly measurable. Then f is Bochner μ-integrable iff $\|f(\cdot)\|$ is μ-integrable. In this case $\left\| \int_S f \, d\mu \right\| \leq \int_S \|f(\cdot)\| \, d\mu$.*

Proof.   ⇒: Let $\{g_j\}$ satisfy (i) and (ii).   Then $\|f(\cdot)\| \leq \|g_j(\cdot)\| + \|f(\cdot) - g_j(\cdot)\|$ implies $\|f(\cdot)\|$ is μ-integrable.

⇐: Let $\{g_j\}$ be a sequence of simple functions such that $g_j \to f$ $\mu - a.e.$ Put $h_j(t) = g_j(t)$ if $\|g_j(t)\| \leq 2\|f(t)\|$ and put $h_j(t) = 0$ otherwise. Then each h_j is a simple function and $h_j \to f$ $\mu - a.e.$ with $\|h_j(\cdot)\| \leq 2\|f(\cdot)\|$. Since

$$\|h_j(\cdot) - f(\cdot)\| \leq 3\|f(\cdot)\|$$

and $\|f(\cdot)\|$ is μ-integrable, the Dominated Convergence Theorem gives

$$\int_S \|h_j(\cdot) - f(\cdot)\|\, d\mu \to 0$$

so f is Bochner μ-integrable.

Also, the Dominated Convergence Theorem implies that

$$\lim \int_S \|h_j(\cdot)\|\, d\mu = \int_S \|f(\cdot)\|\, d\mu$$

so

$$\left\| \int_S f(\cdot) d\mu \right\| = \lim \left\| \int_S h_j(\cdot) d\mu \right\| \leq \lim \int_S \|h_j(\cdot)\|\, d\mu = \int_S \|f(\cdot)\|\, d\mu. \qquad \square$$

We next compare the Bochner and Pettis integrals for strongly measurable functions.

Proposition 5.16. *Let $f : S \to X$ be strongly measurable. If f is Bochner μ-integrable, then f is Pettis μ-integrable and the two integrals agree.*

Proof.   If $x' \in X'$,

$$|x'(f(\cdot))| \leq \|f(\cdot)\|\, \|x'\|$$

so $x'f$ is μ-integrable. In the computations which follow $P \int f d\mu$ $[B \int f d\mu]$ will denote the Pettis integral [Bochner integral]. Let $\{g_j\}$ be as in Definition 5.14. The Dominated Convergence Theorem implies

$$\lim x'\left(P \int_S g_j d\mu\right) = \lim x'\left(B \int_S g_j d\mu\right)$$

$$= x'\left(B \int_S f d\mu\right) = \lim \int_S x' g_j d\mu = \int_S x' f d\mu$$

so f is Pettis integrable with $P \int_S f d\mu = B \int_S f d\mu$. $\qquad \square$

Of course, a function can be Pettis integrable without being strongly measurable; however, the converse of Proposition 5.16 fails even for strongly measurable functions.

Example 5.17. Let μ be Lebesgue measure on $S = [1, \infty)$ and let $\sum_k x_k$ be a series in X which is norm subseries convergent. Define $f : S \to X$ by

$$f = \sum_{k=1}^{\infty} x_k \chi_{[k,k+1)}.$$

Then

$$\int_1^{\infty} x' f d\mu = \sum_{k=1}^{\infty} x'(x_k) = x'(\sum_{k=1}^{\infty} x_k)$$

so f is Pettis μ-integrable with $P \int_1^{\infty} f d\mu = \sum_{k=1}^{\infty} x_k$. Since

$$\|f(\cdot)\| = \sum_{k=1}^{\infty} \|x_k\| \chi_{[k,k+1)},$$

f is Bochner μ-integrable iff the series $\sum_k x_k$ is absolutely convergent. Thus, if we choose a subseries convergent series $\sum_k x_k$ which is not absolutely convergent [e.g., $\sum_k \frac{1}{k} e^k$ in c_0], then f furnishes an example of a strongly measurable function which is Pettis integrable but not Bochner integrable. Actually this example along with the Dvoretsky-Rogers Theorem shows that the Pettis and Bochner integrals coincide with respect to μ for strongly measurable X valued functions iff X is finite dimensional (the Dvoretsky-Rogers Theorem states that X is finite dimensional iff every subseries convergent series is absolutely convergent [Sw1] 30.1).

We next consider basic properties of the indefinite Bochner integral. If $m : \Sigma \to X$ is finitely additive, the variation of m over $A \in \Sigma$ is defined to be

$$var(m)(A) = \sup\{\sum_{j=1}^{n} \|m(A_j)\|\},$$

where the supremum is taken over all partitions $\{A_j\}$ of A with $A_j \in \Sigma$. See [Sw1] (2.2.1.7) for the scalar case. It is readily checked that $var(m)(\cdot)$ is finitely additive when m is finitely additive and is countably additive when m is countably additive.

Theorem 5.18. *Let $f : S \to X$ be Bochner μ-integrable and let $F : \Sigma \to X$ be the indefinite Bochner integral of f, $F(A) = \int_A f d\mu$. Then*

(i) F is absolutely continuous with respect to μ in the sense that for every $\epsilon > 0$ there exists $\delta > 0$ such that $\mu(A) < \delta$ implies $\|F(A)\| < \epsilon$,

(ii) F is norm countably additive,

(iii) F has finite variation with $var(F)(A) = \int_A \|f(\cdot)\| \, d\mu$.

Proof. (i) follows from Theorem 5.15 and since F is obviously finitely additive, (ii) also follows from Theorem 5.15.

(iii): Let $\{A_j : j = 1, ..., n\} \subset \Sigma$ be a partition of A. then by Theorem 5.15,

$$\sum_{j=1}^{n} \|F(A_j)\| \leq \sum_{j=1}^{n} \int_{A_j} \|f(\cdot)\| \, d\mu = \int_A \|f(\cdot)\| \, d\mu$$

so

$$var(F)(A) \leq \int_A \|f(\cdot)\| \, d\mu. \tag{5.1}$$

For the reverse inequality, first suppose $f = \sum_{j=1}^{n} x_j \chi_{A_j}$ is a simple function with the $\{A_j\}$ disjoint and $\bigcup_{j=1}^{n} A_j = S$. Consider the partition $\{A \cap A_j\}$ of A. We have

$$var(F)(A) \geq \sum_{j=1}^{n} \|F(A \cap A_j)\| = \sum_{j=1}^{n} \|x_j\| \, \mu(A \cap A_j) = \int_A \|f(\cdot)\| \, d\mu$$

so we have $var(F)(A) = \int_A \|f(\cdot)\| \, d\mu$ for simple functions. Now suppose f is Bochner μ-integrable and pick simple functions $\{g_j\}$ such that $g_j \to f$ $\mu - a.e.$ with $\|g_j(\cdot)\| \leq 2 \|f(\cdot)\|$ [see the proof of Theorem 5.15]. Set $F_j = \int g_j d\mu$. From (5.1),

$$|var(F_j)(A) - var(F)(A)| \leq var(F_j - F)(A) \leq \int_A \|g_j(\cdot) - f(\cdot)\| \, d\mu \to 0.$$

By the Dominated Convergence Theorem,

$$var(F_j)(A) = \int_A \|g_j(\cdot)\| \, d\mu \to \int_A \|f(\cdot)\| \, d\mu$$

so $\int_A \|f(\cdot)\| \, d\mu = var(F)(A)$. $\square$

Condition (iii) of Theorem 5.18 furnishes a nice test for the Bochner integrability of a Pettis integrable function.

Theorem 5.19. *Let $f : S \to X$ be strongly measurable and Pettis μ-integrable. Set $F(A) = P \int_A f d\mu$. Then f is Bochner μ-integrable iff F has finite variation.*

Proof. $\Leftarrow$: First assume that μ is finite. Let
$$A_k = \{t \in S : k - 1 \le \|f(t)\| < k\}$$
so $\{A_k\}$ is pairwise disjoint with $S = \cup_{k=1}^{\infty} A_k$. Since f is Bochner integrable over each A_k (Theorem 5.15),
$$var(F)(A_k) = \int_{A_k} \|f(\cdot)\| \, d\mu$$
by Theorem 5.18.(iii). Since F is norm countably additive (Theorem 5.12), it is easily checked that $var(F)$ is countably additive. By the Monotone Convergence Theorem,
$$var(F)(S) = \sum_{k=1}^{\infty} var(F)(A_k) = \sum_{k=1}^{\infty} \int_{A_k} \|f(\cdot)\| \, d\mu = \int_S \|f(\cdot)\| \, d\mu$$
so f is Bochner integrable by Theorem 5.15.

If μ is σ-finite, $S = \cup_{j=1}^{\infty} B_j$ where the $\{B_j\}$ are pairwise disjoint and $\mu(B_j) < \infty$. By the part above and the countable additivity of $var(F)$,
$$var(F)(S) = \sum_{j=1}^{\infty} var(F)(B_j) = \sum_{j=1}^{\infty} \int_{B_j} \|f(\cdot)\| \, d\mu = \int_S \|f(\cdot)\| \, d\mu$$
and f is Bochner integrable by Theorem 5.15.

$\Rightarrow$: This follows from Theorem 5.18. $\qquad\square$

Theorem 5.20. *(Dominated Convergence Theorem) Let $f_j : S \to X$ be Bochner μ-integrable for $j \in \mathbb{N}$ and suppose $\{f_j\}$ converges to $f : S \to X$ $\mu - a.e.$ If there exists $g \in L^1(\mu)$ such that*
$$\|f_j(\cdot)\| \le g \; \mu - a.e.,$$
then f is Bochner μ-integrable and
$$\lim \int_S \|f_j(\cdot) - f(\cdot)\| \, d\mu = 0$$
so, in particular,
$$\lim \int_S f_j d\mu = \int_S f d\mu.$$

Proof. Since $\|f_j(\cdot)\| \to \|f(\cdot)\| \; \mu-a.e.$, the Dominated Convergence Theorem implies $\|f(\cdot)\|$ is μ-integrable. Proposition 5.7 and Theorem 5.15 imply f is Bochner μ-integrable. Since $\|f_j(\cdot) - f(\cdot)\| \to 0 \; \mu - a.e.$, and $\|f_j(\cdot) - f(\cdot)\| \le 2g$, the Dominated Convergence Theorem again yields
$$\lim \int_S \|f_j(\cdot) - f(\cdot)\| \, d\mu = 0.$$
Since
$$\left\| \int_S f_j d\mu - \int_S f d\mu \right\| \le \int_S \|f_j(\cdot) - f(\cdot)\| \, d\mu,$$
the last statement also follows. $\qquad\square$

Let $\mathcal{B}(\mu, X)$ be the space of all X valued Bochner μ-integrable functions $f : S \to X$. Then $\mathcal{B}(\mu, X)$ is a vector space and carries a natural semi-norm

$$\|f\|_b = \int_S \|f(\cdot)\| \, d\mu.$$

In contrast to the case of the Pettis integral, we show that $\mathcal{B}(\mu, X)$ is complete under this semi-norm.

Theorem 5.21. $\mathcal{B}(\mu, X)$ *is complete with respect to* $\|\cdot\|_b$.

Proof.   Let $\{f_j\}$ be a Cauchy sequence in $\mathcal{B}(\mu, X)$. Pick a subsequence $\{n_j\}$ such that

$$\left\| f_{n_{j+1}} - f_{n_j} \right\|_b < 1/2^j.$$

By the Monotone Convergence Theorem,

$$\sum_{j=1}^{\infty} \int_S \left\| f_{n_{j+1}}(\cdot) - f_{n_j}(\cdot) \right\| \, d\mu = \int_S \sum_{j=1}^{\infty} \left\| f_{n_{j+1}}(\cdot) - f_{n_j}(\cdot) \right\| \, d\mu < \infty$$

so the series $\sum_{j=1}^{\infty} \left\| f_{n_{j+1}}(\cdot) - f_{n_j}(\cdot) \right\|$ converges $\mu - a.e.$ to a real valued μ-integrable function g. Therefore, the series

$$\sum_{j=1}^{\infty} (f_{n_{j+1}} - f_{n_j}) + f_{n_1} = \lim f_{n_j}$$

converges $\mu - a.e.$ in X. Let $f(t) = \lim f_{n_j}(t)$ when this limit exists and 0 otherwise. If $\epsilon > 0$, then

$$\int_S \left\| f_{n_k}(\cdot) - f_{n_j}(\cdot) \right\| \, d\mu < \epsilon$$

for large j, k so Fatou's Lemma implies

$$\int_S \left\| f(\cdot) - f_{n_j}(\cdot) \right\| \, d\mu \le \epsilon$$

for large j. Hence, $f \in \mathcal{B}(\mu, X)$ and $\left\| f_{n_j} - f \right\|_b \to 0$ so $\|f_j - f\|_b \to 0$. $\square$

The results above show that the Bochner integral enjoys many of the same properties as the Lebesgue integral. In the case of the Lebesgue integral if $\nu : \Sigma \to \mathbb{R}$ is a signed measure with finite variation such that ν is absolutely continuous with respect to μ (in the sense of Theorem 5.18.(i)), then there exists a μ-integrable function $f : S \to \mathbb{R}$ such that $\nu = \int f d\mu$ (Radon–Nikodym Theorem). However, the analogous result is false for the Bochner integral.

Example 5.22. Let $S = [0,1)$ and μ be Lebesgue measure on S. Let $E_1 = [0, 1/2)$, $E_2 = [1/2, 1)$, $E_3 = [0, 1/3)$, $E_4 = [1/3, 2/3)$, $E_5 = [2/3, 1)$, ... and set $f_j = \chi_{E_j}$. Define $\nu : \Sigma \to c_0$ by

$$\nu(A) = \{\mu(A \cap E_j)\}.$$

Now ν is finitely additive and $\|\nu(A)\| \leq \mu(A)$ so ν is countably additive and $var(\nu) \leq \mu$ so ν is absolutely continuous with respect to μ in the sense of Theorem 5.18.(i). Suppose there exists a Bochner μ-integrable $f : S \to c_0$ such that $\nu = \int f d\mu$. Then

$$e^j(\nu(A)) = \mu(A \cap E_j) = \int_A \chi_{E_j} d\mu = \int_A e^j f(t) d\mu(t)$$

so $\chi_{E_j} = e^j(f(\cdot))$ $\mu - a.e.$ and

$$f = \sum_{j=1}^{\infty} \chi_{E_j} e^j \quad \mu - a.e.$$

However,

$$\sum_{j=1}^{\infty} \chi_{E_j}(t) e^j \notin c_0$$

for any $t \in S$. Hence, ν has no Radon-Nikodym derivative in $\mathcal{B}(\mu, X)$ with respect to μ.

Although the vector measure ν in Example 5.22 does not have a Radon–Nikodym derivative or density with values in c_0, the function

$$f = \sum_{j=1}^{\infty} \chi_{E_j} e^j : [0, 1] \to l^\infty$$

is in some sense a density with values outside the space c_0. Thomas has studied this property of densities with values outside the range space of a vector measure ([Th]).

We now collect several special results for vector integrals which will be used in later chapters.

Theorem 5.23. *Let $f : S \to X$, Y be a Banach space and $T \in L(X, Y)$.*

(i) *If f is Dunford integrable, then $Tf = T \circ f$ is Dunford integrable and $T \int_S f d\mu = \int_S T f d\mu$ (Dunford integrals).*

(ii) *If f is Pettis integrable, then Tf is Pettis integrable and $T \int_S f d\mu = \int_S T f d\mu$ (Pettis integrals).*

(iii) If f is Bochner integrable, then Tf is Bochner integrable and $T \int_S f d\mu = \int_S T f d\mu$ (Bochner integrals).

Proof.   (i): Let $y' \in Y'$. Then $y'T = T'y' \in X'$ so $y'Tf$ is integrable and Tf is Dunford integrable with

$$y' \int_S T f d\mu = \int_S y' T f d\mu = \int_S T' y' f d\mu = T' y' \int_S f d\mu$$

and $T \int_S f d\mu = \int_S T f d\mu$ as Dunford integrals.

(ii): $\int_S f d\mu \in X$ so $T \int_S f d\mu \in Y$ and (ii) follows from (i).

(iii): Tf is clearly strongly measurable and $\|Tf(\cdot)\| \le \|T\| \|f(\cdot)\|$ implies Tf is Bochner integrable and equality follows from (ii).   □

We will later need a version of Fubini's Theorem for the Bochner integral. Let Λ be a σ-algebra of subsets of a set T and let ν be a σ-finite measure on Λ. We denote the product measure of μ and ν by $\mu \times \nu$. In what follows below we assume that all functions which are defined almost everywhere are always extended by setting the functions involved to be 0 on the set where they are not defined.

Theorem 5.24. *(Fubini) Suppose $f : S \times T \to X$ is $\mu \times \nu$ Bochner integrable. Then*

(i) $\int_T f(s,t) d\nu(t)$ exists as a Bochner integral for μ-almost all $s \in S$,
(ii) the function $s \to \int_T f(s,t) d\nu(t)$ is Bochner μ-integrable and
(iii) $\int_{S \times T} f d\mu \times \nu = \int_S \int_T f(s,t) d\nu(t) d\mu(s)$.

Proof.   We may assume f is separably valued. For each $x' \in X'$ and $s \in S$, $x'f(s, \cdot)$ is measurable which implies $f(s, \cdot)$ is strongly measurable. Fubini's Theorem applied to $\|f(\cdot, \cdot)\|$ implies

$$\int_T \|f(s,t)\| \, d\nu(t) = \int_T \|f(s, \cdot)\| \, d\nu < \infty$$

for μ almost all s. Therefore,

$$\int_T f(s, \cdot) d\nu = \int_T f(s,t) d\nu(t)$$

exists for μ almost all s. Now

$$x' \int_T f(s, \cdot) d\nu = \int_T x' f(s, \cdot) d\nu$$

implies the function $s \to \int_T x' f(s, \cdot) d\nu$ is measurable and since

$$\int_T f(s, \cdot) d\nu \in \overline{span}\{f(s, t) : s \in S, t \in T\},$$

$s \to \int_T f(s, \cdot) d\nu$ is separably valued and, therefore, strongly measurable. Since

$$\int_S \left\| \int_T f(s, \cdot) d\nu \right\| d\mu(s) \leq \int_S \int_T \|f(s, t)\| \, d\nu(t) d\mu(s)$$

$$= \int_{S \times T} \|f(s, t)\| \, d\mu \times \nu(s, t) < \infty,$$

$s \to \int_T f(s, \cdot) d\nu$ is μ-integrable. Fubini's Theorem applied to $x' f(\cdot, \cdot)$ gives

$$x' \int_{S \times T} f d\mu \times \nu = \int_{S \times T} x' f d\mu \times \nu = \int_S \int_T x' f(s, t) d\nu(t) d\mu(s)$$

so

$$\int_{S \times T} f d\mu \times \nu = \int_S \int_T f(s, t) d\nu(t) d\mu(s). \qquad \square$$

We will also need an integration by parts formula.

Definition 5.25. Let $f : [a, b] \to X$ and $t \in [a, b]$. The function f is differentiable at t if the limit,

$$\lim_{h \to 0} \frac{f(t + h) - f(t)}{h} = v$$

exists; the value v is called the derivative of f at t and is denoted by $f'(t)$.

We have the following simple version of the Fundamental Theorem of Calculus.

Proposition 5.26. If $f : [a, b] \to X$ is differenitiable at all points of $[a, b]$ and f' is continuous on $[a, b]$, then

$$\int_a^b f' d\lambda = f(b) - f(a),$$

where λ is Lebesgue measure and the integral is a Bochner integral.

Proof. Since f' is strongly measurable and bounded it is clear that f' is Bochner integrable. If $x' \in X'$,

$$x' \int_a^b f' d\lambda = \int_a^b x' f' d\lambda = x'(f(b) - f(a))$$

by classical results so $\int_a^b f' d\lambda = f(b) - f(a)$. $\qquad \square$

We can now establish a simple version of an integration by parts formula.

Proposition 5.27. *(Integration by Parts) Let $f : [a, b] \to X, g : [a, b] \to \mathbb{C}$ both be differentiable on $[a, b]$ with continuous derivatives. Then*

$$\int_a^b gf'd\lambda + \int_a^b g'fd\lambda = g(b)f(b) - g(a)f(a).$$

Proof. The product differentiation rule $(gf)' = g'f + gf'$ holds so the result follows from Proposition 5.26. $\square$

Remark 5.28. For the history and a more complete development of the Pettis and Bochner integrals, see [DU].

Chapter 6

An Abstract Functional Calculus

In this chapter we will introduce the axioms for a functional calculus and give an example of such a calculus. In later chapters there will be further examples given. Let $\mathcal{F}$ be an algebra of either real or complex valued functions defined on a subset of either $\mathbb{R}$ or $\mathbb{C}$ and let $\mathcal{P}$ be the algebra of either real or complex polynomials depending on whether $\mathcal{F}$ contains either real or complex valued functions. Let X be a topological vector space and $\mathcal{K}$ an algebra of continuous linear operators from X into itself with the product as composition.

Definition 6.1. An abstract functional calculus on $\mathcal{K}$ relative to $\mathcal{F}$ is a map $\Theta : \mathcal{F} \times \mathcal{K} \to \mathcal{K}$ such that

> (1) $\Theta(1, \mathbf{A}) = I$ for every $\mathbf{A} \in \mathcal{K}$,
> (2) $\Theta(t, \mathbf{A}) = \mathbf{A}$ for every $\mathbf{A} \in \mathcal{K}$,
> (3) $\Theta(\cdot, \mathbf{A})$ is an algebra homomorphism for every $\mathbf{A} \in \mathcal{K}$.

It follows immediately from the axioms that $\Theta(p, \mathbf{A}) = p(\mathbf{A})$ for every $\mathbf{A} \in \mathcal{K}$, $p \in \mathcal{P}$.

Definition 6.2. If $\mathcal{F}$ is closed under composition and

$$\Theta(f \circ g, \mathbf{A}) = \Theta(f, \Theta(g, \mathbf{A}))$$

for every $f, g \in \mathcal{F}$, $\mathbf{A} \in \mathcal{K}$, then the functional calculus Θ is said to be complete.

For example, if $\mathcal{F} = \mathcal{P}$ and $\Theta(p, \mathbf{A}) = p(\mathbf{A})$, then Θ is complete.

Theorem 6.3. *Assume that $\mathcal{F}$ and $\mathcal{K}$ are Frechet algebras with $\mathcal{P}$ dense in $\mathcal{F}$. Let Θ be an abstract functional calculus on $\mathcal{K}$ relative to $\mathcal{P}$.*

(1) If $\Theta(\cdot, \mathbf{A})$ is continuous for every $\mathbf{A} \in \mathcal{K}$, then Θ has a unique extension, $\overline{\Theta}$, to an abstract functional calculus on $\mathcal{K}$ relative to $\mathcal{F}$ such that $\overline{\Theta}(\cdot, \mathbf{A})$ is continuous for every $\mathbf{A} \in \mathcal{K}$.

(2) If the bilinear map $(f, g) \to f \circ g$ from $\mathcal{F} \times \mathcal{F}$ into $\mathcal{F}$ is separately continuous and $\Theta(p, \cdot)$ is continuous from $\mathcal{K}$ into $\mathcal{K}$ for every $p \in \mathcal{P}$, then the extension $\overline{\Theta}$ is complete.

(3) If the map $\Theta(\cdot, \mathbf{A}) : \mathcal{P} \to \mathcal{P}$ $(\Theta(p, \mathbf{A}) = p(\mathbf{A}))$ is uniformly continuous when $\mathbf{A}$ runs over bounded subsets of $\mathcal{K}$, then $\overline{\Theta}(f, \cdot) : \mathcal{K} \to \mathcal{K}$ is continuous for every $f \in \mathcal{F}$.

Proof. Since $\mathcal{K}$ is complete, every $\Theta(\cdot, \mathbf{A})$ has a unique continuous linear extension to $\mathcal{F}$ so (1) follows.

(2): Let $f, g \in \mathcal{F}$ and pick $p_j, q_j \in \mathcal{P}$ such that $p_j \to f, q_j \to g$. Then $p_j \circ q_k(\mathbf{A}) = p_j(q_k(\mathbf{A}))$ implies

$$\Theta(p_j \circ q_k, \mathbf{A}) = \Theta(p_j, \Theta(q_k(\mathbf{A}))).$$

Fix j. Then $p_j \circ q_k \to p_j \circ g$ as $k \to \infty$ by hypothesis. By continuity of $\Theta(\cdot, \mathbf{A})$,

$$\Theta(p_j \circ q_k, \mathbf{A}) \to \overline{\Theta}(p_j \circ g, \mathbf{A}).$$

Also $\Theta(q_k, \mathbf{A}) \to \overline{\Theta}(g, \mathbf{A})$ so

$$\Theta(p_j, \Theta(q_k(\mathbf{A}))) \to \Theta(p_j, \overline{\Theta}(g, \mathbf{A}))$$

by hypothesis ($\Theta(p, \cdot)$ is continuous). Therefore,

$$\overline{\Theta}(p_j \circ g, \mathbf{A}) = \Theta(p_j, \overline{\Theta}(g, \mathbf{A})).$$

Since $p_j \circ g \to f \circ g$ by hypothesis,

$$\overline{\Theta}(p_j \circ g, \mathbf{A}) \to \overline{\Theta}(f \circ g, \mathbf{A}).$$

Also, $\Theta(p_j, \overline{\Theta}(g, \mathbf{A})) \to \overline{\Theta}(f, \overline{\Theta}(g, \mathbf{A}))$ so

$$\overline{\Theta}(f, \overline{\Theta}(g, \mathbf{A})) = \overline{\Theta}(f \circ g, \mathbf{A}).$$

(3): First observe that $\overline{\Theta}(f, \cdot)$ is uniformly continuous when $\mathbf{A}$ runs over bounded subsets of $\mathcal{F}$. Let $f \in \mathcal{F}$. Pick $p_j \in \mathcal{P}$ such that $p_j \to f$ in $\mathcal{F}$. Let $\mathbf{A} \in \mathcal{K}$ and $\mathbf{A}_j \in \mathcal{K}$, $\mathbf{A}_j \to \mathbf{A}$. Let $\|\cdot\|$ be a continuous semi-norm on $\mathcal{K}$. Then

$$\begin{aligned}
\left\|\overline{\Theta}(f, \mathbf{A}_j) - \overline{\Theta}(f, \mathbf{A})\right\| \leq{}& \left\|\overline{\Theta}(f, \mathbf{A}_j) - \overline{\Theta}(p_k, \mathbf{A}_j)\right\| + \\
& \left\|\overline{\Theta}(p_k, \mathbf{A}_j) - \overline{\Theta}(p_k, \mathbf{A})\right\| + \\
& \left\|\overline{\Theta}(p_k, \mathbf{A}) - \overline{\Theta}(f, \mathbf{A})\right\|.
\end{aligned} \tag{6.1}$$

Let $\epsilon > 0$. By uniform continuity there exists k_1 such that

$$\left\| \overline{\Theta}(f, \mathbf{A}_j) - \overline{\Theta}(p_k, \mathbf{A}_j) \right\| < \epsilon$$

for $k \geq k_1, j \in \mathbb{N}$. Since $\overline{\Theta}(\cdot, \mathbf{A})$ is continuous, there exists $k_2 > k_1$ such that

$$\left\| \overline{\Theta}(p_k, \mathbf{A}) - \overline{\Theta}(f, \mathbf{A}) \right\| < \epsilon$$

for $k \geq k_2$. Now $\Theta(p_{k_2}, \cdot)$ is continuous so there exists j_0 such that

$$\left\| \overline{\Theta}(p_{k_2}, \mathbf{A}_j) - \overline{\Theta}(p_{k_2}, \mathbf{A}) \right\| < \epsilon$$

for $j \geq j_0$. From (6.1)

$$\left\| \overline{\Theta}(f, \mathbf{A}_j) - \overline{\Theta}(f, A) \right\| < 3\epsilon$$

for $j \geq j_0$. Therefore, $\overline{\Theta}(f, \mathbf{A}_j) \to \overline{\Theta}(f, \mathbf{A})$. $\qquad\square$

We give an example of an abstract functional calculus. Further examples are given in later chapters. Let $\mathcal{A}$ be a Banach algebra of commuting self adjoint operators on a Hilbert space H which contains the identity operator I. Let $\mathcal{F} = \mathcal{C}(\mathbb{R})$ be the Frechet algebra of real valued, continuous functions on $\mathbb{R}$ with the topology of uniform convergence on compact subsets of $\mathbb{R}$ and $\mathcal{P}$ the algebra of real polynomials which is dense in $C(\mathbb{R})$ by the Weierstrass Approximation Theorem.

Theorem 6.4. *There exists a complete abstract functional calculus* Θ *on* $\mathcal{A}$ *relative to* $\mathcal{C}(\mathbb{R})$ *which is continuous in each variable.*

Proof. For $\mathbf{A} \in \mathcal{A}$ the map $p \to p(\mathbf{A})$ is continuous from $\mathcal{P} \to \mathbf{A}$ since

$$\|p(\mathbf{A})\| \leq \sup\{|p(t)| : |t| \leq \|\mathbf{A}\|\}$$

(Theorems B.19 and B.23). Therefore, $\Theta(p, \mathbf{A}) = p(\mathbf{A})$ is continuous in the first variable and from (1) of Theorem 6.3, Θ has a unique extension to an abstract functional calculus on $\mathcal{A}$ relative to $\mathcal{C}(\mathbb{R})$, which we still denote by Θ, and which is continuous in the first variable.

To complete the proof, we verify conditions (2) and (3) of Theorem 6.3.

(2): First we show $(f, g) \to f \circ g$ from $\mathcal{C}(\mathbb{R}) \times \mathcal{C}(\mathbb{R}) \to \mathcal{C}(\mathbb{R})$ is separately continuous. Suppose $f_j \to f$ in $\mathcal{C}(\mathbb{R})$ and $g \in \mathcal{C}(\mathbb{R})$. Let $K \subset \mathbb{R}$ be compact. Then

$$\sup_{K}\{|f_j \circ g - f \circ g|\} = \sup_{g(K)}\{|f_j - f|\} \to 0.$$

Next, suppose $f \in \mathcal{C}(\mathbb{R})$ and $g_j \to g$ in $\mathcal{C}(\mathbb{R})$. Let $\epsilon > 0$. There exists n_0 such that $|g_j(t) - g(t)| < 1$ for $t \in K$, $j \geq n_0$. Set $K_1 = \{t : dist(t, g(K)) \leq 1\}$

and note $K_1 \supset g(K)$ is compact since g is continuous. f is uniformly continuous on K_1 so there exists $\delta > 0$ such that

$$|f(t) - f(s)| < \epsilon \quad \text{if} \quad s, t \in K_1, \ |s - t| < \delta.$$

If $j \geq n_0, t \in K$, then $g_j(t) \in K_1$. There exists $n_1 > n_0$ such that

$$|g_j(t) - g(t)| < \delta \quad \text{for} \quad j \geq n_1, \ t \in K.$$

Therefore, if $j \geq n_1, t \in K$, then $g_j(t) \in K_1, g(t) \in K_1$ and $|g_j(t) - g(t)| < \delta$ so

$$|f(g_j(t)) - f(g(t))| < \epsilon$$

and $f \circ g_j \to f \circ g$ in $\mathcal{C}(\mathbb{R})$.

Next, we show $\mathbf{A} \to p(\mathbf{A})$ is continuous from $\mathcal{A}$ into $\mathcal{A}$ for any $p \in \mathcal{P}$. For this it suffices to consider the case when $p(t) = t^k$. If $\mathbf{A}, \mathbf{B} \in \mathcal{A}$, then

$$\mathbf{A}^k - \mathbf{B}^k = \sum_{j=0}^{k-1} \mathbf{A}^{k-j-1}(\mathbf{A} - \mathbf{B})\mathbf{B}^j$$

so if $\mathbf{A}_i \to \mathbf{A}$ in $\mathcal{A}$ and $\max\{\|\mathbf{A}_j\|, \|\mathbf{A}\|\} < a$, then

$$\left\| \mathbf{A}^k - \mathbf{A}_i^k \right\| \leq \sum_{j=0}^{k-1} \|\mathbf{A}\|^{k-j-1} \|\mathbf{A} - \mathbf{A}_i\| \|\mathbf{A}_i\|^j \leq k a^{k-1} \|\mathbf{A} - \mathbf{A}_i\| \to 0.$$

It follows from (2) of Theorem 6.3 that Θ is complete.

(3): Suppose $\mathcal{B} \subset \mathcal{A}$ is bounded with $\|\mathbf{A}\| \leq a$ for every $\mathbf{A} \in \mathcal{B}$. Then

$$\|p(\mathbf{A})\| \leq \sup\{|p(t)| : |t| \leq \|\mathbf{A}\|\} \leq \sup\{|p(t)| \leq a\}$$

for $\mathbf{A} \in \mathcal{B}$. If $p_j \to p$ in $\mathcal{P}$, then $\{p_j\}$ converges uniformly to p on $[-a, a]$ so

$$\|p_j(\mathbf{A}) - p(\mathbf{A})\| \leq \sup\{|p_j(t) - p(t)| : |t| \leq a\} \to 0$$

uniformly for $\mathbf{A} \in \mathcal{B}$.

It follows from (3) of Theorem 6.3 that Θ is continuous in the second variable.

We have shown that $\Theta : \mathcal{C}(\mathbb{R}) \times \mathcal{A} \to \mathcal{A}$ defined above is a complete functional calculus on $\mathcal{A}$, a Banach algebra of commuting operators, which is continuous in each variable separately. Actually, $\Theta : \mathcal{C}(\mathbb{R}) \times \mathcal{A} \to \mathcal{A}$ is jointly continuous in this case as we now show. Let $f_n \to f$ in $\mathcal{C}(\mathbb{R})$ and $\mathbf{A}_n \to \mathbf{A}$ in $\mathcal{A}$. Then

$$\begin{aligned}
\|\Theta(f_n, \mathbf{A}_n) - \Theta(f, \mathbf{A})\| &= \|f_n(\mathbf{A}_n) - f(\mathbf{A})\| \\
&\leq \|f_n(\mathbf{A}_n) - f(\mathbf{A}_n)\| + \|f(\mathbf{A}_n) - f(\mathbf{A})\| \\
&= \|(f_n - f)(\mathbf{A}_n)\| + \|f(\mathbf{A}_n) - f(\mathbf{A})\|.
\end{aligned}$$

Let $M > 0$ be such that $\|\mathbf{A}_n\| \leq M$ for all n. Then

$$\|(f_n - f)(\mathbf{A}_n)\| \leq \sup_{|t| \leq \|\mathbf{A}_n\|} |f_n(t) - f(t)| \leq \sup_{|t| \leq M} |f_n(t) - f(t)| \to 0$$

since $f_n \to f$ in $\mathcal{C}(\mathbb{R})$. Since Θ is continuous in the first variable, $f_n(\mathbf{A}) \to f(\mathbf{A})$ and it follows that $\Theta(f_n, \mathbf{A}_n) - \Theta(f, \mathbf{A}) \to 0$. $\qquad\square$

Generally speaking there are two different but closely connected ways to construct a functional calculus.

Procedure (1): The homomorphism Θ is defined on some dense subalgebra $\mathcal{F}_0$ of $\mathcal{F}$, an algebra which should at least contain an algebra of smooth functions such as polynomials or $C_c^\infty(\Omega)$, the algebra of infinitely differentiable functions on $\Omega \subset \mathbb{R}^n$ with compact support, depending on the spectrum of the elements in $\mathcal{K}$, where $\mathcal{K}$ may be an algebra of n-tuples of self adjoint operators on a complex Hilbert space. $\mathcal{F}_0$ should have a simple structure (for example, polynomials) which will allow a simple definition of Θ. After that Θ extends to $\mathcal{F}$ by continuity.

Procedure (2): Consider a reproducing formula for functions from $\mathcal{F}$ of the form

$$f(x) = \int k(x, y) f(y) dy.$$

For example, a Cauchy type formula or a Fourier type transform. If the kernel $k(\cdot, \cdot)$ can be extended in a natural way to an operator $k(\mathbf{A}, y)$, then we define the functional calculus by the formula

$$\Theta(f, \mathbf{A}) = f(\mathbf{A}) = \int k(\mathbf{A}, y) f(y) dy.$$

The first procedure is sometimes easier, but the second procedure gives an explicit formula for the functional calculus. Sometimes it is possible to construct the functional calculus in both ways.

An example of Procedure (1) is given in the example above (Theorem 6.4). In Chapter 7 we will use a very similar procedure and in Chapter 8 we will use Procedure (2).

Chapter 7

The Riesz Operational Calculus

In this chapter we offer an exposition of the Riesz Operational Calculus. The calculus was introduced by F. Riesz ([Ri]) and further developed by N. Dunford ([Du1], [Du2]). The calculus is based on Cauchy's integral formula from complex analysis and requires properties of vector valued analytic functions. We begin this chapter with an exposition of the properties of vector valued analytic functions which will be required in the exposition.

Let X be a complex Banach space, D an open subset of $\mathbb{C}$, $f : D \to X$ and $z_0 \in D$. As was the case with integrating vector valued functions, in Chapter 5, there are two choices for defining analyticity of vector valued functions, a strong definition and a weak definition. However in contrast to the situation in integration where these two approaches led to different integrals, the Bochner and Pettis integrals, Theorem 7.2 below will show that the two approaches lead to the same concept of analyticity.

Definition 7.1. f is strongly differentiable at z_0 with derivative $f'(z_0) \in X$ if

$$\lim_{z \to z_0} \left\| \frac{f(z) - f(z_0)}{z - z_0} - f'(z_0) \right\| = 0.$$

f is strongly analytic on D if $f'(z)$ exist for every $z \in D$. The function f is weakly analytic on D if $x'f$ is analytic on D for every $x' \in X'$.

We have the remarkable result of Dunford that these two notions of analyticity coincide.

Theorem 7.2 (Dunford). *f is strongly analytic on D iff f is weakly analytic on D.*

Proof. If f is strongly analytic, it is clearly weakly analytic.

Suppose f is weakly analytic on D. Let $z_0 \in D$. It suffices to show

$$\frac{f(z) - f(z_0)}{z - z_0} - \frac{f(w) - f(z_0)}{w - z_0} \to 0$$

in norm as w, $z \to z_0$ (Cauchy Criterion). Let C be a positively oriented circle with center at z_0 and radius r such that C and its interior are contained in D. For $x' \in X'$, $x'f$ is continuous on D and, therefore, bounded on C. By the Uniform Boundedness Principle, f is norm bounded on C. Let $M > 0$ be such that $\|f(z)\| \le M$ for $z \in C$. Suppose

$$0 < |z - z_0| < \frac{r}{2}, \ 0 < |w - z_0| < \frac{r}{2}.$$

By Cauchy's Integral Formula,

$$x' f(z_0) = \frac{1}{2\pi i} \int_C \frac{x'f(u)}{u - z_0} du,$$

$$x' f(z) = \frac{1}{2\pi i} \int_C \frac{x'f(u)}{u - z} du,$$

$$x' f(w) = \frac{1}{2\pi i} \int_C \frac{x'f(u)}{u - w} du.$$

Then if $\|x'\| \le 1$,

$$\left| \left\langle x', \frac{f(z) - f(z_0)}{z - z_0} - \frac{f(w) - f(z_0)}{w - z_0} \right\rangle \right| =$$

$$\left| \frac{1}{2\pi i} \int_C x'f(u) \frac{z - w}{(u - z)(u - w)(u - z_0)} du \right|$$

$$\le \frac{M|z - w|}{2\pi \frac{r}{2} \frac{r}{2} r} 2\pi r = \frac{4M|z - w|}{r^2}$$

since

$$|u - z| \ge \frac{r}{2}, \ |u - w| \ge \frac{r}{2}$$

for $u \in C$. Therefore,

$$\left\| \frac{f(z) - f(z_0)}{z - z_0} - \frac{f(w) - f(z_0)}{w - z_0} \right\| \le \frac{4M|z - w|}{r^2}$$

for

$$0 < |z - z_0| < \frac{r}{2}, \ 0 < |w - z_0| < \frac{r}{2}. \qquad \square$$

Remark 7.3. Note the theorem implies that the difference quotients
$$\frac{f(z) - f(z_0)}{z - z_0}$$
converge in norm as $z \to z_0$ iff the quotients converge weakly — a remarkable result somewhat analogous to the Orlicz-Pettis Theorem (see Theorem A.6).

Henceforth, we will use the term analytic instead of strongly or weakly analytic.

Note also that if f is analytic on D, then f has (strong) derivatives of all orders on D.

Many of the results in classical complex analysis carry over to vector valued analytic functions by replacing norms by absolute values or by using Theorem 7.2 on the equivalence of strong and weak analyticity. In particular, if C is a piecewise continuous, rectifiable curve contained in D and $f : D \to X$ is continuous, the (Riemann-Stieltjes) integral of f over C,
$$\int_C f(z)dz,$$
is defined exactly as in classical complex analysis. The completeness of X insures the convergence of the Riemann sums defining the integral. Moreover, it is easily seen that if $\mathbf{A} \in L(X)$ or $x' \in X'$, then
$$\mathbf{A} \int_C f(z)dz = \int_C \mathbf{A} f(z)dz,$$
$$\left\langle x', \int_C f(z)dz \right\rangle = \int_C x' f(z)dz.$$

The Cauchy integral formula for analytic functions carries over to vector valued analytic functions.

Theorem 7.4 (Cauchy Integral Formula). *Let $f : D \to X$ be analytic, $z \in D$ and C be a simple closed positively oriented, rectifiable curve whose interior lies in D with z in the interior of C. Then*
$$f(z) = \frac{1}{2\pi i} \int_C \frac{f(w)}{w - z} dw.$$

Proof. The function f is continuous on C so the integral
$$\int_C \frac{f(w)}{w - z} dw$$

exists. For $x' \in X'$,

$$\left\langle x', \frac{1}{2\pi i} \int_C \frac{f(w)}{w-z} dw \right\rangle = \frac{1}{2\pi i} \int_C \frac{x' f(w)}{w-z} dw = x' f(z)$$

by the classical Cauchy Integral Formula for scalar valued analytic functions. Since X' separates the points of X, this gives the result. $\square$

As noted after Theorem 7.2 an analytic function has derivatives of all orders, and we have

$$\frac{d^m}{dz^m} x' f(z) = x' f^{(m)}(z).$$

So using the proof in Theorem 7.4 we also have

$$f^{(m)}(z) = \frac{m!}{2\pi i} \int_C \frac{f(w)}{(w-z)^{m+1}} dw \text{ for } z \in D, \ m = 0, 1, 2, \dots . \qquad (7.1)$$

Using Theorem 7.2 we have a converse of Theorem 7.4

Theorem 7.5. *Suppose $f : D \to X$ is continuous. If the Cauchy integral formula holds for f inside D, then f is analytic in D.*

Proof.   Simply apply the scalar version of this theorem to $x' f$ for $x' \in X$ and then apply Theorem 7.2. $\square$

Corollary 7.6 (Analyticity for Continuous Functions). *Let $f : D \to X$ be continuous and suppose $Z \subset X'$ separates the points of X. Then f is analytic on D iff $x' f$ is analytic on D for every $x' \in Z$.*

Proof.   If f is analytic on D, the result is clear.

Suppose $x' f$ is analytic on D for $x' \in Z$. Let C be a simple closed positively oriented, rectifiable curve in D with z in the interior of C and the interior of C inside D. The integral

$$\int_C \frac{f(w)}{w-z} dw$$

exists since f is continuous and for $x' \in Z$,

$$\left\langle x', \frac{1}{2\pi i} \int_C \frac{f(w)}{w-z} dw \right\rangle = \frac{1}{2\pi i} \int_C \frac{x' f(w)}{w-z} dw = x' f(z)$$

by the scalar Cauchy integral formula. Then

$$f(z) = \frac{1}{2\pi i} \int_C \frac{f(w)}{w-z} dw$$

since Z separates points. f is analytic on D by Theorem 7.5. $\square$

Corollary 7.7. *Let Y be a complex Banach space, $f : D \to L(X,Y)$ continuous. Then f is analytic on D iff $\langle y', f(\cdot)x \rangle$ is analytic in D for every $x \in X$, $y' \in Y'$.*

Proof. The result follows from Corollary 7.6 since the set
$$Z = \{y' \otimes x : x \in X, \ y' \in Y'\} \subset L(X,Y)'$$
separates the points of $L(X,Y)$, where we have that $\langle y' \otimes x, \ \mathbf{T} \rangle = \langle y', \mathbf{T}x \rangle$, $\mathbf{T} \in L(X,Y)$. $\square$

We now give several examples where Theorem 7.2 can be used to derive analogues of scalar results in complex analysis for vector valued analytic functions.

Theorem 7.8 (Cauchy's Theorem). *Let $f : D \to X$ be analytic in D and C be a simple closed rectifiable curve whose interior lies in D. Then*
$$\int_C f(z)dz = 0.$$

Proof. If $x' \in X'$, then
$$\left\langle x', \int_C f(z)dz \right\rangle = \int_C x' f(z)dz = 0$$
by the scalar version of Cauchy's Theorem. Since X' separates the points of X, the result follows. $\square$

Theorem 7.9 (Maximum Principle). *Suppose D is connected, $f : D \to X$ is analytic in D and $\|f(\cdot)\|$ is not constant. Then $\|f(\cdot)\|$ cannot attain an absolute maximum at any point of D.*

Proof. Suppose there exists $z_0 \in D$ such that
$$\|f(z)\| \le \|f(z_0)\|$$
for $z \in D$. Pick $x' \in X'$, $\|x'\| = 1$, such that
$$\langle x', \ f(z_0) \rangle = \|f(z_0)\|.$$
Then $x'f$ is analytic in D and
$$|\langle x', \ f(z) \rangle| \le \|f(z)\| \le \|f(z_0)\| = \langle x', \ f(z_0) \rangle.$$
The scalar Maximum Principle implies $x'f$ is constant with value $\|f(z_0)\|$. But $|\langle x', \ f(z) \rangle| \le \|f(z)\|$ and there exists a $z \in D$ with $\|f(z)\| < \|f(z_0)\|$ so we have the desired contradiction. $\square$

Theorem 7.10 (Liouville). *Let $f : \mathbb{C} \to X$ be analytic (entire). If f is bounded, then f is a constant function.*

Proof. For each $x' \in X'$, $x'f$ is entire and bounded and, therefore, constant by Louiville's Theorem. Hence, $\langle x', f(z) \rangle = \langle x', f(0) \rangle$ for all $z \in \mathbb{C}$ and $f(z) = f(0)$ for $z \in \mathbb{C}$.   □

As an application of Liouville's Theorem, we give a proof of Fuglede's Theorem on the commutativity of a normal operator and its adjoint.

Lemma 7.11. *Let $\mathbf{S} \in L(H)$, where H is a complex Hilbert space. Then*
$$\left\| e^{\mathbf{S} - \mathbf{S}^*} \right\| = 1 .$$

Proof. Set $\mathbf{V} = \mathbf{S} - \mathbf{S}^*$. Define
$$\mathbf{Q} = e^{\mathbf{V}} = \sum_{j=0}^{\infty} \frac{\mathbf{V}^j}{j!} .$$
Since
$$\mathbf{V} = -\mathbf{V}^*, \mathbf{Q}^* = e^{V^*} = e^{-\mathbf{V}} = \mathbf{Q}^{-1}$$
and $\mathbf{Q}$ is unitary so
$$\|\mathbf{Q}\| = 1 = \left\| e^{\mathbf{S} - \mathbf{S}^*} \right\| .$$ □

Theorem 7.12. *Let $\mathbf{M}, \mathbf{N}, \mathbf{T} \in L(H)$ with $\mathbf{M}, \mathbf{N}$ normal. If $\mathbf{MT} = \mathbf{TN}$, then $\mathbf{M}^*\mathbf{T} = \mathbf{TN}^*$.*

Proof. By induction $\mathbf{M}^j\mathbf{T} = \mathbf{TN}^j$ for $j \in \mathbb{N}$ so $e^{\mathbf{M}}\mathbf{T} = \mathbf{T}e^{\mathbf{N}}$ and
$$\mathbf{T} = e^{-\mathbf{M}}\mathbf{T}e^{\mathbf{N}}. \tag{7.2}$$
Put $\mathbf{U}_1 = e^{\mathbf{M}^* - \mathbf{M}}$ and $\mathbf{U}_2 = e^{\mathbf{N} - \mathbf{N}^*}$. Since $\mathbf{M}$ and $\mathbf{N}$ are normal, (7.2) implies $e^{\mathbf{M}^*}\mathbf{T}e^{-\mathbf{N}^*} = \mathbf{U}_1\mathbf{T}\mathbf{U}_2$. By Lemma 7.11
$$\|\mathbf{U}_1\| = \|\mathbf{U}_2\| = 1$$
so
$$\left\| e^{\mathbf{M}^*}\mathbf{T}e^{-\mathbf{N}^*} \right\| \le \|\mathbf{T}\| . \tag{7.3}$$
Define
$$f(z) = e^{z\mathbf{M}^*}\mathbf{T}e^{-z\mathbf{N}^*} , \quad z \in \mathbb{C} . \tag{7.4}$$
The hypothesis of the theorem holds with $\bar{z}\mathbf{M}$, $\bar{z}\mathbf{N}$ replacing $\mathbf{M}$, $\mathbf{N}$ so (7.3) implies $\|f(z)\| \le \|\mathbf{T}\|$ for $z \in \mathbb{C}$. Therefore, f is a bounded entire function and $f(z) = f(0) = \mathbf{T}$ for all $z \in \mathbb{C}$ by Liouville's Theorem. Now (7.4) implies $e^{z\mathbf{M}^*}\mathbf{T} = \mathbf{T}e^{z\mathbf{N}^*}$ for all $z \in \mathbb{C}$. Differentiating this expression and setting $z = 0$ gives $\mathbf{M}^*T = TN^*$.   □

Corollary 7.13 (Fuglede). *If $\mathbf{N}$ is normal and commutes with $\mathbf{T} \in L(H)$, then $\mathbf{N}^*$ commutes with $\mathbf{T}$.*

7.1 Power Series

As is the case in classical complex analysis, every analytic function has a unique power series expansion about any point in its domain. We can derive this from the formulas for the derivatives given in equation (7.1). Suppose f is analytic in D and $z_0 \in D$. Let r_0 be the radius of a circle C with center z_0 such that C and its interior is contained in D. Since

$$f^m(z_0) = \frac{m!}{2\pi i} \int_C \frac{f(w)}{(w-z)^{m+1}} dw, \quad \left\| f^{(m)}(z_0) \right\| \le \frac{m!M}{r_0^m},$$

where

$$M = \sup \{ \|f(w)\| : \|z_0 - w\| = r_0 \}.$$

If $0 < r_1 < r_0$, then this estimate implies the power series

$$\sum_{m=0}^{\infty} (z - z_0)^m \frac{f^{(m)}(z_0)}{m!}$$

converges uniformly and absolutely for $|z - z_0| \le r_1$. Since

$$\frac{d^m}{dz^m} x' f(z) = x' f^{(m)}(z)$$

and

$$x' f(z) = \sum_{m=0}^{\infty} \frac{(z - z_0)^m}{m!} \frac{d^m}{dz^m} x' f(z_0),$$

this series converges to $f(z)$ for $|z - z_0| \le r_1$. This also shows that the coefficients in the power series expansion are unique.

7.2 Laurent Series

Similar to the case of power series expansion for vector valued analytic functions, functions with isolated singularities have Laurent Expansions. Fix z_0. For $r < s$ let

$$A(r, s) = \{ z : r \le |z - z_0| \le s \}$$

be the annulus determine by r, s, z_0 and let

$$C_r = \{ z : |z - z_0| = r \}$$

be the circle with center z_0, radius r. Suppose $f : D \to X$ is analytic in D and $D \supset A(r, s)$. Let

$$M_r = \max \{ \|f(z)\| : \|z - z_0\| = r \}.$$

Let

$$a_n = \frac{1}{2\pi i} \int_{C_r} \frac{f(w)}{(w - z_0)^{n+1}} dw, \ n \in \mathbb{N} \cup \{0\},$$

$$b_m = \frac{1}{2\pi i} \int_{C_s} \frac{f(w)}{(w - z_0)^{-m+1}} dw, \ m \in \mathbb{N}.$$

Consider the (Laurent) series

$$\sum_{n=0}^{\infty} a_n(z - z_0)^n + \sum_{n=1}^{\infty} \frac{b_n}{(z - z_0)^n}.$$

As in case of the power series, we have the estimates

$$\|a_n\| \le \frac{M_s}{s^n}, \ \|b_n\| \le \frac{M_r}{r^{-n}}.$$

Suppose $r < r_1 < s_1 < s$. As before for power series, the series

$$\sum_{n=0}^{\infty} a_n(z - z_0)^n$$

converges uniformly and absolutely for $|z - z_0| \le s_1$. For the series

$$\sum_{n=1}^{\infty} \frac{b_n}{(z - z_0)^n},$$

for $|z - z_0| \ge r_1$, we have

$$\frac{\|b_n\|}{|z - z_0|^n} \le M_r (\frac{r}{r_1})^n$$

so the series $\sum_{n=1}^{\infty} \frac{b_n}{(z-z_0)^n}$ converges absolutely and uniformly for $r_1 \le |z - z_0| \le s_1$.

The fact that the Laurent series converges to f and the coefficients are unique follow by considering the functions $x'f$ as above for power series.

7.3 Runge's Theorem

Runge's Theorem is important in both classical and functional analysis. The classical proof of translation of poles is somewhat tedious and long [SZ]. On the other side the abstract proof needs a consequence of the Hahn-Banach Theorem and sometimes the Riesz Representation Theorem [Ru1], [RT].

The proof given here is elementary and is due to S. Grabiner [Gr]. It is based on the three following lemmas. But first let's write down Runge's Theorem.

Theorem 7.14 (Runge). *Let K be a compact subset of $\mathbb{C}$ and E a subset of $(\mathbb{C} \cup \{\infty\}) \backslash K$ that contains at least one point in each component of $(\mathbb{C} \cup \{\infty\}) \backslash K$. If f is analytic in an open set containing K, then there are rational functions r_n whose poles lie in E such that r_n converges uniformly to f in K.*

Lemma 7.15. *Every analytic function on an open set containing K is the uniform limit on K of rational functions, all of whose poles lie in the complement of K.*

Lemma 7.16. *If V and U are open subsets in $\mathbb{C}$ with $V \subset U$ and with the boundary of V disjoint from U, then V contains every component of U which it meets.*

Lemma 7.17. *Let $B(E) = \{f : f$ is a uniform limit on K of rational functions whose poles lie in $E\}$. If $\lambda \notin K$, then $(z - \lambda)^{-1}$ belongs to $B(E)$.*

Proof. [Runge's Theorem] Since $B(E)$ is an algebra closed under uniform limits (a Banach algebra), Runge's theorem is an immediate consequence of Lemmas 7.15 and 7.17, Lemma 7.16 is used to prove Lemma 7.17. $\square$

Note that in the special case where $(\mathbb{C} \cup \{\infty\}) \backslash K$ is connected we can take $E = \{\infty\}$ and the sequence of rational functions will actually be a sequence of polynomials.

Proof. [Lemma 7.15] Let D be an open set containing K. Using the Cauchy integral theorem there exists a smooth, simple, closed, positively oriented, rectifiable curve Γ from $[0, 1]$ into $D \backslash K$ such that for every analytic function f and $z \in K$.

$$f(z) = \frac{1}{2\pi i} \int_\Gamma \frac{f(w)}{w - z} dw.$$

Take $\varepsilon > 0$ and $\delta = dist(\Gamma^*, K)$, where Γ^* denotes the image of $[0, 1]$, the domain of Γ; $\delta > 0$ since Γ^* and K are disjoint and compact. Let

$s, t \in [0, 1]$, $z \in K$. Then

$$\left| \frac{f(\Gamma(t))}{\Gamma(t) - z} - \frac{f(\Gamma(s))}{\Gamma(s) - z} \right| = \left| \frac{f(\Gamma(t))(\Gamma(s) - z) - f(\Gamma(s))(\Gamma(t) - z)}{(\Gamma(t) - z)(\Gamma(s) - z)} \right|$$

$$= \left| \frac{f(\Gamma(t))(\Gamma(s) - \Gamma(t)) + \Gamma(t)(f(\Gamma(t)) - f(\Gamma(s))) - z(f(\Gamma(t)) - f(\Gamma(s)))}{(\Gamma(t) - z)(\Gamma(s) - z)} \right|$$

$$\leq \frac{1}{\delta^2} \Big(|f(\Gamma(t))| \, |\Gamma(s) - \Gamma(t)| + |\Gamma(t)| \, |f(\Gamma(t)) - f(\Gamma(s))|$$

$$+ |z| \, |f(\Gamma(t)) - f(\Gamma(s))| \Big).$$

Now Γ and $f \circ \Gamma$ are bounded functions and K is a compact set so there is a $C > 0$ such that the previous expression is bounded by

$$\frac{C}{\delta^2} \left(|\Gamma(s) - \Gamma(t)| + 2 \, |f(\Gamma(t)) - f(\Gamma(s))| \right).$$

Γ and $f \circ \Gamma$ are uniformly continuous so there exists on the interval $[0, 1]$ a partition $0 = t_0 < t_1 < ... < t_n = 1$ such that for $t \in [t_{j-1}, t_j]$ and $z \in K$ we have

$$|\Gamma(t) - \Gamma(t_j)| < \frac{\varepsilon \delta^2}{2C}$$

and

$$|f(\Gamma(t)) - f(\Gamma(t_j))| < \frac{\varepsilon \delta^2}{4C}.$$

Then

$$\left| \frac{f(\Gamma(t))}{\Gamma(t) - z} - \frac{f(\Gamma(t_j))}{\Gamma(t_j) - z} \right| < \varepsilon.$$

Define

$$R(z) = \sum_{j=1}^{n} \frac{f(\Gamma(t_j))}{\Gamma(t_j) - z} (\Gamma(t_j) - \Gamma(t_{j-1})),$$

where clearly $z \neq \Gamma(t_j)$.

Then $R(z)$ is a rational function where poles are in the set $\{\Gamma(t_1), \Gamma(t_2), ..., \Gamma(t_n)\}$; in particular the poles are in $D \setminus K$. For all $z \in K$

we have

$$
|2\pi i f(z) - R(z)| = \left| \int_\Gamma \frac{f(w)}{w - z} dw - \sum_{j=1}^{n} \frac{f(\Gamma(t_j))}{\Gamma(t_j) - z} (\Gamma(t_j) - \Gamma(t_j - 1)) \right|
$$

$$
= \left| \sum_{j=1}^{n} \int_{t_{j-1}}^{t_j} \left(\frac{f(\Gamma(t))}{\Gamma(t) - z} - \frac{f(\Gamma(t_j))}{\Gamma(t_j) - z} \right) \Gamma'(t) dt \right|
$$

$$
\leq \varepsilon \left| \sum_{j=1}^{n} \int_{t_{j-1}}^{t_j} \Gamma'(t) dt \right|
$$

$$
\leq \varepsilon \int_0^1 |\Gamma'(t)| \, dt
$$

$$
= \varepsilon \cdot \text{length of } \Gamma
$$

and since the length of Γ is independent of ε, Lemma 7.15 is proved. $\square$

Proof. [Lemma 7.16] Suppose H is a component of U which contains a point s in V. Let G be the component of V that contains s. Since H is connected, it must be either equal to G or contain a boundary point of G, but each boundary point of G is a boundary point of V and cannot be in U. $\square$

Proof. [Lemma 7.17] For large enough $|\lambda|$, the Taylor series for $(z - \lambda)^{-1}$ converges uniformly on K. Suppose $\infty \in E$. Thus $(z - \lambda)^{-1} \in B(E)$ and $B((E \setminus \{\infty\}) \cup \{\lambda\}) \subseteq B(E)$.

Suppose $f \in B((E \setminus \{\infty\}) \cup \{\lambda\})$ and R a rational function with poles in $(E - \{\infty\}) \cup \{\lambda\}$ that approximates f. Write $R = R_1 + R_2$ where all poles (if any) of R, are in $E \setminus \{\infty\}$ and the pole (if any) of R_2 is at λ. But R_2 can be approximated by a polynomial P; hence, $R_1 + P$ approximates f and has its poles in E, so $f \in B(E)$. Thus, it is sufficient to establish the lemma for sets $E \subset \mathbb{C}$. Now put $U = \mathbb{C} \setminus K$ and define $V = \{\lambda \in U : (z - \lambda)^{-1} \in B(E)\}$. By hypothesis we have that $E \subset U$ and hence $E \subset V \subset U$. To use Lemma 7.16 we need first to prove that V is open. Let $\lambda \in V$ and take μ such that $0 < |\lambda - \mu| < dist(\lambda, K)$. Then $\mu \in \mathbb{C} \setminus K$ and for all $z \in K$:

$$
\frac{1}{z - \mu} = \frac{1}{(z - \lambda)(1 - \frac{\mu - \lambda}{z - \lambda})}
$$

which is a uniformly convergent sum of a series in powers of $(z - \lambda)^{-1}$. Thus, $\mu \in V$ and this proves V is open. Now we will prove the boundary of

V and U are disjoint. Take μ in the boundary of V and choose a sequence $\{\lambda_n\}$ in V with limit μ. Since μ does not belong to V we must have

$$|\lambda_n - \mu| \geq dist(\lambda_n, K)$$

for all $n \in \mathbb{N}$. Since $|\lambda_n - \mu| \to 0$ the distance from μ to K must be 0 and so $\mu \in K$. Then $\mu \notin U$ and the boundary of V and U are disjoint. Hence, the hypothesis of Lemma 7.16 holds. Let H be any component of U. By definition of E, there exists $s \in E$ such that $s \in H$. Now $s \in V$ since $E \subset V$, thus $H \cap V \neq \emptyset$ and Lemma 7.16 implies $H \subset V$. We have shown that every component of V is a subset of V, consequently $U \subset V$ and since $V \subset U$ we have $U = V$. $\qquad\qquad\square$

7.4   Several Complex Variables

In the last chapter we will also need results for analytic functions of several complex variables. We record these results here.

Definition 7.18. Let $D \subset \mathbb{C}^n$ be open and $f : D \to X$. The function is analytic in D if for each j, the function $f(z_1, ..., z_{j-1}, \cdot, z_{j+1}, ..., z_n)$ is analytic. That is, if f is analytic in each variable separately.

Remark 7.19. If $f : D \to X$ is C^∞, then f is analytic on D iff the Cauchy-Riemann equations $\frac{\partial f}{\partial \bar{z}_j} = 0$ for every j, where

$$\frac{\partial}{\partial \bar{z}_j} = \frac{1}{2}\left\{\frac{\partial f}{\partial x_j} + i\frac{\partial f}{\partial y_j}\right\}, \quad z_j = x_j + iy_j.$$

We will need the Cauchy integral formula for analytic functions of several variables. For this it is convenient to have an important result due to Hartog.

Theorem 7.20 (Hartog). *If f is analytic in D, then f is jointly continuous, i.e., continuous in all variables.*

Hartog's theorem is surprisingly difficult to prove. For a proof, see [Be]. We say that f is analytic on a subset $E \subset \mathbb{C}^n$ if there is an open set $D \supset E$ such that f is analytic on D.

Theorem 7.21. (*Cauchy Integral Formula*) Let $r_j > 0$ for $j = 1, ..., n$, $z^0 = (z_1^0, ..., z_n^0)$, and

$$P = \left\{z = (z_1, ..., z_n) : \left|z_j - z_j^0\right| \leq r_j, \ j = 1, ..., n\right\}.$$

If f is analytic on P, then

$$f(z_1, ..., z_n) = (\frac{1}{2\pi i})^n \int_{|z_1 - z_1^0| = r_1} ... \int_{|z_n - z_n^0| = r_n} \prod_{j=1}^{n} \frac{1}{w_j - z_j} f(w_1, ..., w_n) dw_1 ... dw_n.$$

Proof. $(n = 2)$ Assume $z^0 = (0, 0)$ and $f(\cdot, z_2)$ is analytic for $|z_2| \leq r_2$ so by the Cauchy Integral Formula

$$f(z_1, z_2) = \frac{1}{2\pi i} \int_{|z_1| = r_1} \frac{1}{w_1 - z_1} f(w_1, w_2) dw_1.$$

Similarly, $f(z_1, \cdot)$ is analytic for $|z_1| \leq r_1$ so by the Cauchy Integral Formula

$$f(z_1, z_2) = \frac{1}{2\pi i} \int_{|z_1| = r_1} \frac{1}{w_1 - z_1} \left\{ \frac{1}{2\pi i} \int_{|z_2| = r_2} \frac{1}{w_2 - z_2} f(w_1, w_2) dw_2 \right\} dw_1.$$

By Hartog's Theorem, we can write this iterated integral as a double integral. $\qquad \square$

Corollary 7.22. *An analytic function $f : D \to X$ has derivatives of all orders, i.e., f is C^∞.*

This follows by differentiating under the integral sign.

Corollary 7.23. *If $f : D \to X$ is analytic, $\frac{\partial f}{\partial z_j}$ is analytic for every j.*

We will also need the following result.

Theorem 7.24. *If $f_k : D \to X$ is analytic on D for each $k \in \mathbb{N}$ and $f_k \to f$ uniformly on compact subsets of D, then f is analytic on D.*

For a history of vector valued analytic functions in functional analysis, see [Ta2].

Example 7.25. An elementary and powerful example that will be used later is the exponential function. For every $N \in \mathbb{N}$, and $\mathbf{A}_i \in L(X)$, where X is a complex Banach space,

$$p_N(\xi_1, ..., \xi_n) = \sum_{k=0}^{N} \frac{(2\pi i)^k}{k!} (\xi_1 \mathbf{A}_1 + ... + \xi_n \mathbf{A}_n)^k,$$

is a polynomial in $(\xi_1, ..., \xi_n)$ and so it is an entire function. As $N \to \infty$, $p_N(\xi_1, ..., \xi_n)$ converges uniformly on compact sets to $e^{2\pi i \xi \mathbf{A}}$, so it is also an entire function.

7.5   Riesz Operational Calculus

The Riesz operational calculus which we now describe is based on the properties of vector valued analytic functions. The calculus is based on the Cauchy integral formula for analytic functions f,

$$f(z) = \frac{1}{2\pi i} \int_C \frac{f(w)}{w - z} dw = \frac{1}{2\pi i} \int_C f(w)(w - z)^{-1} dw.$$

If $\mathbf{A} \in L(X)$, to define $f(\mathbf{A})$ for an analytic function we consider replacing $(w - z)^{-1}$ by $(w - \mathbf{A})^{-1} = R_w(\mathbf{A})$, the resolvent operator, so

$$f(\mathbf{A}) = \frac{1}{2\pi i} \int_C f(w)(w - \mathbf{A})^{-1} dw = \frac{1}{2\pi i} \int_C f(w) R_w(\mathbf{A}) dw.$$

If f is analytic on C, then f is continuous and $R_w(\mathbf{A})$ is bounded on C so the integral defining $f(\mathbf{A})$ exists as a Bochner integral. This operational calculus was first introduced by F. Riesz ([Ri]) and the basic properties were further developed by N. Dunford in ([Du1], [Du2]). We now give the details of the Riesz operational calculus. Throughout the remainder of this chapter $\mathbf{A}$ will be an operator in $L(X)$.

Definition 7.26. Let $\Im(\mathbf{A})$ be the set of all functions f which are analytic on an open neighborhood U of $\sigma(\mathbf{A})$ [the neighborhood U may depend on f].

By a Jordan curve C we mean a finite number $C_1, ..., C_k$ of disjoint, simple, closed, positively oriented, rectifiable curves. If $f \in \Im(\mathbf{A})$ is analytic on an open neighborhood V of $\sigma(\mathbf{A})$, we say that V is *admissible* (for V and $\mathbf{A}$) if $\sigma(\mathbf{A}) \subset V \subset \overline{V} \subset U$ and ∂V, the boundary of V, is a Jordan curve.

It is intuitive that given $\sigma(\mathbf{A})$ and U, there exists a admissible neighborhood. However, the proof of the existence is somewhat complicated. We refer the reader to [Ta1] for the existence.

If f is an analytic function on an open neighborhood of U containing $\sigma(\mathbf{A})$, we define

$$f(\mathbf{A}) = \frac{1}{2\pi i} \int_{\partial V} f(z) R_z(\mathbf{A}) dz,$$

where V is any admissible domain with ∂V positively oriented.

Note that the integral defining $f(\mathbf{A})$ is independent of the choice of the admissible domain V by Cauchy's Theorem.

Theorem 7.27. *Let f, $g \in \Im(\mathbf{A})$ and s, $t \in \mathbb{C}$. Then*

(1) $sf + tg \in \Im(\mathbf{A})$ *and* $(sf + tg)(\mathbf{A}) = sf(\mathbf{A}) + tg(\mathbf{A})$,

(2) $f \cdot g \in \Im(\mathbf{A})$ *and* $f(\mathbf{A})g(\mathbf{A}) = (fg)(\mathbf{A})$,

(3) *if* $f \equiv 1$, *then* $f(\mathbf{A}) = I$,

(4) *if* $f(z) = z$ *for all* $z \in \mathbb{C}$, *then* $f(\mathbf{A}) = \mathbf{A}$,

(5) *if* $f(z) = z^k$ *for* $k = 0, 1, 2, ...,$*then* $f(\mathbf{A}) = \mathbf{A}^k$,

(6) *if* f *has a Maclaurin expansion* $f(z) = \sum\limits_{k=0}^{\infty} a_k z^k$ *in a neighborhood*

of $\sigma(\mathbf{A})$, *then* $f(\mathbf{A}) = \sum\limits_{k=0}^{\infty} a_k \mathbf{A}^k$ *(norm convergence)*,

(7) $f \in \Im(\mathbf{A}')$ *and* $f(\mathbf{A}') = f(\mathbf{A})$, *where* $\mathbf{A}'$ *denote the transpose or adjoint of* $\mathbf{A}$.

Proof.

(1) is clear.

(2) Clearly $f, g \in \Im(\mathbf{A})$. Choose two open neighborhoods U_1, U_2 of $\sigma(\mathbf{A})$ such that U_i is admissible and such that $\sigma(\mathbf{A}) \subset U_1 \subset \overline{U}_1 \subset U_2$ and $f.g$ are analytic in $\overline{U_2}$. Then

$$
f(\mathbf{A})g(\mathbf{A}) = -\frac{1}{4\pi^2} \Big(\int\limits_{\partial U_1} f(z) R_z(\mathbf{A}) dz \Big) \Big(\int\limits_{\partial U_2} g(w) R_w(\mathbf{A}) dw \Big)
$$

$$
= -\frac{1}{4\pi^2} \int\limits_{\partial U_1} \int\limits_{\partial U_2} f(z)g(w) R_z(\mathbf{A}) R_w(\mathbf{A}) dw dz
$$

since $\int\limits_{\partial U_2} f(w) R_w(\mathbf{A}) dw$ is a bounded, linear operator.

By the resolvent equation and the Cauchy integral formula,

$$
f(\mathbf{A})g(\mathbf{A}) = -\frac{1}{4\pi^2} \int\limits_{\partial U_1} \int\limits_{\partial U_2} f(z)g(w) \frac{R_z(\mathbf{A}) - R_w(\mathbf{A})}{w - z} dw dz
$$

$$
= -\frac{1}{4\pi^2} \int\limits_{\partial U_1} f(z) \int\limits_{\partial U_2} \left\{ \frac{g(w)}{w - z} R_z(\mathbf{A}) - \frac{g(w)}{w - z} R_w(\mathbf{A}) \right\} dw dz
$$

$$
= -\frac{1}{4\pi^2} \int\limits_{\partial U_1} f(z) 2\pi i g(z) R_z(\mathbf{A}) dz
$$

$$
+ \frac{1}{4\pi^2} \int\limits_{\partial U_1} f(z) \int\limits_{\partial U_2} \frac{g(w)}{w - z} R_w(\mathbf{A}) dw dz.
$$

Since the integrand in the last integral is continuous, we may change the order of integration to obtain

$$f(\mathbf{A})g(\mathbf{A}) = \frac{1}{2\pi i} \int_{\partial U_1} f(z)g(z)R_z(\mathbf{A})dz$$

$$+ \frac{1}{4\pi^2} \int_{\partial U_2} g(w)R_w(\mathbf{A}) \int_{\partial U_1} \frac{f(z)}{w-z} dz dw$$

$$= \frac{1}{2\pi i} \int_{\partial U_1} f(z)g(z)R_z(\mathbf{A})dz + 0$$

$$= (fg)(\mathbf{A}),$$

where we have used Cauchy's Theorem.

(3) If $f \equiv 1$,

$$f(\mathbf{A}) = \frac{1}{2\pi i} \int_C R_z(\mathbf{A})dz,$$

where C is a circle with center at the origin and radius $R > \|A\|$. Then

$$f(A) = \frac{1}{2\pi i} \int_C R_z(\mathbf{A})dz$$

$$= \frac{1}{2\pi i} \int_C (z - \mathbf{A})^{-1} dz$$

$$= \frac{1}{2\pi i} \int_C \frac{1}{z}(I - \frac{\mathbf{A}}{z})^{-1} dz$$

$$= \frac{1}{2\pi i} \int_C \frac{1}{z} \sum_{k=0}^{\infty} \frac{\mathbf{A}^k}{z^k} dz.$$

Since $\left\|\frac{\mathbf{A}}{z}\right\| < 1$ for $|z| = R$, the series in the last integral converges uniformly over C, and we have

$$f(\mathbf{A}) = \sum_{k=0}^{\infty} \frac{1}{2\pi i} \int_C \frac{\mathbf{A}^k}{z^{k+1}} dz = I,$$

since $\int_C \frac{1}{z^{k+1}} dz = 0$ if $k > 1$ and $\int_C \frac{1}{z^{k+1}} dz = 2\pi i$ if $k = 0$.

(4) If $f(z) = z$ for all z, then

$$f(\mathbf{A}) = \frac{1}{2\pi i} \int_C z R_z(\mathbf{A})dz$$

as in (3), and we have $f(\mathbf{A}) = \mathbf{A}$ by the same proof.

(5) This follows from (2), (3) and (4).

(6) The power series for f converges uniformly on a circle C with center 0 and radius $R > r(\mathbf{A})$. From (5),

$$
\begin{aligned}
f(\mathbf{A}) &= \frac{1}{2\pi i} \int_C f(z) R_z(\mathbf{A}) dz \\
&= \frac{1}{2\pi i} \int_C \left(\sum_{k=0}^{\infty} a_k z^k \right) R_z(\mathbf{A}) dz \\
&= \frac{1}{2\pi i} \sum_{k=0}^{\infty} a_k \int_C z^k R_z(\mathbf{A}) dz \\
&= \sum_{k=0}^{\infty} a_k \mathbf{A}^k.
\end{aligned}
$$

(7) This follows since $\sigma(\mathbf{A}) = \sigma(\mathbf{A}^*)$ and $R_z(\mathbf{A})^* = R_z(\mathbf{A}^*)$. $\qquad\square$

It follows from Theorem 7.27 that if p is a complex polynomial, $p(z) = \sum_{k=0}^{n} a_k z^k$, then

$$
p(\mathbf{A}) = \sum_{k=0}^{n} a_k \mathbf{A}^k,
$$

a property which should be satisfied by any functional calculus.

Theorem 7.27 can be useful in computing inverses.

Corollary 7.28. *Let $f \in \Im(\mathbf{A})$. Then $f(\mathbf{A})$ is invertible iff $f(z) \neq 0$ for $z \in \sigma(\mathbf{A})$. The inverse is $g(\mathbf{A})$, where g is any member of $\Im(\mathbf{A})$ such that g is equal to $\frac{1}{f}$ on a neighborhood of $\sigma(\mathbf{A})$.*

Proof. If $\frac{1}{f}$ is analytic on a neighborhood of $\sigma(\mathbf{A})$, then $1 = f \cdot \left(\frac{1}{f} \right)$ on a neighborhood of $\sigma(\mathbf{A})$ so

$$
I = f(\mathbf{A})(\frac{1}{f})(\mathbf{A})
$$

and $f(\mathbf{A})$ is invertible. If $f(z) = 0$ for some $z \in \sigma(\mathbf{A})$, then $f(w) = (w - z)g(w)$, where g is analytic on a neighborhood of $\sigma(\mathbf{A})$. By Theorem 7.27

$$
f(\mathbf{A}) = (\mathbf{A} - z)g(\mathbf{A})
$$

so

$$
I = (\mathbf{A} - z)[g(\mathbf{A})f(\mathbf{A})^{-1}]
$$

which implies $z \in \rho(\mathbf{A})$. $\qquad\square$

We also have an important commutativity result.

Theorem 7.29. *Let $f \in \Im(\mathbf{A})$. Then $f(\mathbf{A})$ commutes with any operator $\mathbf{B} \in L(X)$ which commutes with $\mathbf{A}$.*

Proof. Let C be the boundary of an admissible neighborhood of $\sigma(\mathbf{A})$ so

$$f(\mathbf{A}) = \frac{1}{2\pi i} \int_C f(z) R_z(\mathbf{A}) dz.$$

Since $\mathbf{B}$ commutes with the resolvent $R_z(\mathbf{A})$ by Theorem B.12,

$$\mathbf{B}f(\mathbf{A}) = \frac{1}{2\pi i} \mathbf{B} \int_C f(z) R_z(\mathbf{A}) dz = \frac{1}{2\pi i} \int_C f(z) \mathbf{B} R_z(\mathbf{A}) dz$$

$$= \frac{1}{2\pi i} \int_C f(z) R_z(\mathbf{A}) \mathbf{B} dz$$

$$= \frac{1}{2\pi i} \int_C f(z) R_z(\mathbf{A}) dz \mathbf{B}$$

$$= f(\mathbf{A})\mathbf{B}. \qquad \qquad \square$$

We can use the operational calculus to give a proof of a version of the spectral mapping theorem.

Theorem 7.30. *(Spectral Mapping Theorem)* *If $f \in \Im(\mathbf{A})$, then $f(\sigma(\mathbf{A})) = \sigma(f(\mathbf{A}))$.*

Proof. Let $z \in \sigma(\mathbf{A})$. Define g on the domain of f by

$$g(w) = \begin{cases} \frac{f(z)-f(w)}{z-w} & z \neq w \\ f'(z) & z = w \, . \end{cases}$$

By Theorem 7.27

$$f(z)I - f(\mathbf{A}) = (zI - \mathbf{A})g(\mathbf{A})$$

since $g \in \Im(\mathbf{A})$.

Therefore, if $f(z)I - f(\mathbf{A})$ has a bounded inverse $\mathbf{B} \in L(X)$, $g(\mathbf{A})\mathbf{B}$ will be a bounded inverse in $L(X)$ for $z - \mathbf{A}$. Hence, $f(z) \in \sigma(f(\mathbf{A}))$ and $f(\sigma(\mathbf{A})) \subset \sigma(f(\mathbf{A}))$.

Suppose $w \in \sigma(f(\mathbf{A}))$ and $w \notin f(\sigma(\mathbf{A}))$. Then

$$h(z) = \frac{1}{(f(z) - w)}$$

defines an element of $\Im(\mathbf{A})$. By Theorem 7.27

$$h(\mathbf{A})(f(\mathbf{A}) - wI) = I$$

so $w \notin \sigma(f(\mathbf{A}))$, a contradiction. Hence, $w \in f(\sigma(\mathbf{A}))$ and $\sigma(f(\mathbf{A})) \subset f(\sigma(\mathbf{A}))$. □

We also have a composition theorem for the Riesz operational calculus.

Theorem 7.31. *Let $f \in \Im(\mathbf{A})$, $g \in \Im(f(\mathbf{A}))$ and $h = g \circ f$. Then $h \in \Im(\mathbf{A})$ and $h(\mathbf{A}) = g(f(\mathbf{A}))$.*

Proof. That $h \in \Im(\mathbf{A})$ follows from the Spectral Mapping Theorem 7.30. Let U be an admissible neighborhood of $\sigma(f(\mathbf{A}))$ and suppose g is analytic on a neighborhood of $U \cup \partial U$. Let V be an admissible neighborhood of $\sigma(\mathbf{A})$ such that $\partial V \cup V$ contains a neighborhood where f is analytic. Suppose further that $f(\partial V \cup V) \subset U$. By Theorem 7.27, the operator

$$\mathbf{T}(z) = \frac{1}{2\pi i} \int_{\partial V} \frac{R_w(\mathbf{A})}{z - f(w)} dw$$

satisfies the equation

$$(zI - f(\mathbf{A}))\mathbf{T}(z) = \mathbf{T}(z)(zI - f(\mathbf{A})) = I.$$

Thus,

$$\mathbf{T}(z) = R_z(f(\mathbf{A}))$$

and

$$
\begin{aligned}
g(f(\mathbf{A})) &= \frac{1}{2\pi i} \int_{\partial U} g(z) R_z(f(\mathbf{A})) dz \\
&= -\frac{1}{4\pi^2} \int_{\partial U} \int_{\partial V} \frac{g(z) R_z(\mathbf{A})}{z - f(w)} dw dz \\
&= \frac{1}{2\pi i} \int_{\partial V} R_w(\mathbf{A}) g(f(w)) dw \\
&= h(\mathbf{A}),
\end{aligned}
$$

where the change of the order of integration is valid since the integrand is continuous. □

We next consider continuity properties for the mapping $f(\mathbf{A})$. First, we consider continuity in the variable f.

Theorem 7.32. *Let f, $f_k \in \Im(\mathbf{A})$ and suppose each f_k is analytic on an open neighborhood U of $\sigma(\mathbf{A})$. If $\{f_k\}$ converges to f uniformly on compact subsets of U, then $\|f_k(\mathbf{A}) - f(\mathbf{A})\| \to 0$.*

Proof. Let V be an open neighborhood of $\sigma(\mathbf{A})$ such that $V \subset U$ is admissible. Then $R_z(\mathbf{A})$ is bounded for $z \in \partial V$ so

$$\| f_k(z) R_z(\mathbf{A}) - f(z) R_z(\mathbf{A}) \| \to 0$$

uniformly for $z \in \partial V$ and

$$\| f_z(\mathbf{A}) - f(\mathbf{A}) \| = \frac{1}{2\pi} \left\| \int_{\partial V} (f_k(z) - f(z)) R_z(\mathbf{A}) dz \right\|$$

$$\leq \frac{1}{2\pi} \int_{\partial V} \| (f_k(z) - f(z)) R_z(\mathbf{A}) \| \, d\,|z| \to 0. \qquad \square$$

The map $\theta : \Im(\mathbf{A}) \to L(X)$, $\theta(f) = f(\mathbf{A})$, has many of the properties of a homomorphism (Theorem 7.27), but the set $\Im(\mathbf{A})$ is not a vector space or an algebra. However, we can define an equivalence relation on $\Im(\mathbf{A})$ which converts $\Im(\mathbf{A})$ into an algebra and such that the induced map from θ is a homomorphism. If f, $g \in \Im(\mathbf{A})$, we say that $f \sim g$ iff f and g agree on a neighborhood of $\sigma(\mathbf{A})$. The relation $f \sim g$ clearly defines an equivalence relation on $\Im(\mathbf{A})$, and if $f \sim g$, then $\theta(f) = \theta(g)$. As is often the case we do not distinguish between the elements $f \in \Im(\mathbf{A})$ and the equivalence class determined by $\sim$. In this case the induced map θ is an algebra homomorphism from the equivalence classes of $\Im(\mathbf{A})$ into $L(X)$ which satisfies the continuity property of Theorem 7.32. We now establish a uniqueness result for the operational calculus θ.

Theorem 7.33. *Suppose $\psi : \Im(\mathbf{A}) \to L(X)$ is an algebra homomorphism satisfying*

(1) $\psi(1) = I$
(2) $\psi(z) = \mathbf{A}$
(3) *If $\{f_n\}$ is a sequence of functions analytic on an open neighborhood D of $\sigma(\mathbf{A})$ such that $f_k \to f$ uniformly on compact subsets of D, then $\psi(f_j) \to \psi(f)$.*
   Then $\psi(f) = f(\mathbf{A})$ for every $f \in \Im(\mathbf{A})$.

Proof. First, we show $\psi(f) = f(\mathbf{A})$ when f is a rational function in $\Im(\mathbf{A})$ and then we will apply Runge's Theorem. If

$$m \geq 1, \quad \psi(z^m) = \psi(z)^m = \mathbf{A}^m$$

so for any polynomial

$$\psi(p) = p(\mathbf{A}).$$

Let q be any polynomial such that $\frac{1}{q} \in \Im(\mathbf{A})$. Then

$$I = \psi(1) = \psi(q\frac{1}{q}) = \psi(q)\psi(\frac{1}{q}) = q(\mathbf{A})\psi(\frac{1}{q})$$

so

$$\psi(\frac{1}{q}) = (q(\mathbf{A}))^{-1}$$

and $q(\mathbf{A})$ is invertible. If p, q are polynomials such that $\frac{1}{q} \in \Im(\mathbf{A})$, then

$$\psi(\frac{p}{q}) = \psi(p)\psi(\frac{1}{q}) = p(\mathbf{A})(q(\mathbf{A}))^{-1} = \frac{p}{q}(\mathbf{A})$$

by Theorem 7.27.

Let $f \in \Im(\mathbf{A})$. By Runge's Theorem 7.14 there exists a sequence of rational functions $\{r_n\} \subset \Im(\mathbf{A})$ such that $r_n \to f$ uniformly on compact subsets of a neighborhood of $\sigma(\mathbf{A})$. By (3), $\psi(r_n) \to \psi(f)$. But, $\psi(r_n) = r_n(\mathbf{A})$ so $\psi(f) = f(\mathbf{A})$ by Theorem 7.32. □

Concerning analyticity, we have:

Theorem 7.34. *Let U be a neighborhood of $\sigma(\mathbf{A})$ and $V \subset \mathbb{C}$ open. Suppose $f : U \times V \to X$ is analytic in each variable. Then $f(\mathbf{A}, \cdot) : V \to L(X)$ is analytic on U.*

Proof. Let $W \supset \sigma(\mathbf{A})$ be an admissible open neighborhood of $\sigma(\mathbf{A})$, $\sigma(\mathbf{A}) \subset W \subset \overline{W} \subset U$ and set $C = \partial W$ so

$$f(\mathbf{A}, v) = \frac{1}{2\pi i}\int\limits_C f(z, v)R_z(\mathbf{A})dz \text{ for } v \in V.$$

Fix $v \in V$. We show $f(\mathbf{A}, \cdot)$ is differentiable at v. Pick $r > 0$ such that $\{w : |v - w| \le r\} \subset V$ and let C_r be the circle with center at v and radius r. Let $|h| < r$, $|k| < r$ and consider the difference quotients

$$Q = \frac{f(z, v+h) - f(z, v)}{h} - \frac{f(z, v+k) - f(z)}{k}.$$

By Cauchy's Integral formula,

$$Q = \frac{1}{2\pi i}\int\limits_{C_r} f(z, w)\left\{\frac{\frac{1}{w-v-h} - \frac{1}{w-v}}{h} - \frac{\frac{1}{w-v-k} - \frac{1}{w-v}}{k}\right\} dw$$

$$= \frac{1}{2\pi i}\int\limits_{C_r} f(z, w)\left\{\frac{h - k}{(w - v - h)(w - v - k)(w - v)}\right\} dw.$$

If

$$M = \sup \{|f(z,w)| : z \in C, \ |w - v| \le r\}$$

($M < \infty$ since f is continuous in both variables by Hartog's Theorem 7.20), then

$$|Q| \le M \frac{|h - k|}{(r - |h|)(r - |k|)}.$$

Therefore, the difference quotients Q satisfy a uniform Cauchy condition for $z \in C$. Letting $k \to 0$ means there exists $\delta > 0$ such that

$$\left| \frac{f(z, \ v + h) - f(z, \ v)}{h} - \frac{\partial f}{\partial v}(z, v) \right| < \varepsilon$$

for $|h| < \delta$, $z \in C$. Now $\frac{\partial f}{\partial v}(\cdot, \ w) \in \Im(\mathbf{A})$ for $w \in V$ and

$$\left\| \frac{f(\mathbf{A}, v + h) - f(\mathbf{A}, \ v)}{h} - \left(\frac{\partial f}{\partial v}(\cdot, v) \right)(\mathbf{A}) \right\|$$

$$= \left\| \frac{1}{2\pi i} \int_C \left\{ \frac{f(z, v + h) - f(z, v)}{h} - \frac{\partial f}{\partial v}(z, v) \right\} R_z(\mathbf{A}) dz \right\|$$

$$\le \frac{1}{2\pi} \varepsilon \sup \{\|R_z(\mathbf{A})\| : z \in C\} \ell(C),$$

where $\ell(C)$ is the length of C. Therefore, $f(\mathbf{A}, \ \cdot)$ is differentiable at v with

$$\frac{\partial f}{\partial v}(\mathbf{A}, v) = \left(\frac{\partial f}{\partial v}(\cdot, v) \right)(\mathbf{A}).$$

$\square$

We now consider continuity in the second variable, $\mathbf{A}$, for the mapping $f(\mathbf{A})$. For this we first have a perturbation result.

Theorem 7.35. *(Perturbation) Let $\varepsilon > 0$. There exists $\delta > 0$ such that if $\mathbf{A}_1 \in L(X)$ and $\|\mathbf{A} - \mathbf{A}_1\| < \delta$, then*

$$\sigma(\mathbf{A}_1) \subset \{z : dist(z, \sigma(\mathbf{A})) < \varepsilon\} = S(\sigma(\mathbf{A}), \varepsilon)$$

and

$$\|R_z(\mathbf{A}_1) - R_z(\mathbf{A})\| < \varepsilon$$

for $z \notin S(\sigma(\mathbf{A}), \varepsilon)$.

Proof. Since $\lim\limits_{|z|\to\infty} \|R_z(\mathbf{A})\| = 0$ (Theorem B.13), there exists N such that $\|R_z(\mathbf{A})\| \le N$ for $z \notin S\left(\sigma(\mathbf{A}), \varepsilon\right)$. If $\delta_1 = \frac{1}{N} > \|\mathbf{A}_1 - \mathbf{A}\|$, then for $z \notin S\left(\sigma(\mathbf{A}), \varepsilon\right)$,

$$\|(z - \mathbf{A}) - (z - \mathbf{A}_1)\| \le \frac{1}{N} \le \frac{1}{\|R_z(\mathbf{A})\|}$$

which implies $(z - \mathbf{A}_1)^{-1} \in L(X)$ (Theorem B.8). Hence, $\sigma(\mathbf{A}_1) \subset S(\sigma(\mathbf{A}), \varepsilon)$. Also by Theorem B.8

$$\begin{aligned}
\|R_z(\mathbf{A}_1) - R_z(\mathbf{A})\| &\le \frac{\|R_z(\mathbf{A})\|^2 \|\mathbf{A} - \mathbf{A}_1\|}{(1 - \|R_z(\mathbf{A})\| \|\mathbf{A} - \mathbf{A}_1\|)} \\
&\le \frac{N^2 \|\mathbf{A} - \mathbf{A}_1\|}{(1 - N \|\mathbf{A} - \mathbf{A}_1\|)}
\end{aligned}$$

so that $\|R_z(\mathbf{A}_1) - R_z(\mathbf{A})\| < \varepsilon$ if $\|\mathbf{A} - \mathbf{A}_1\| < \delta_2 = \frac{\varepsilon}{(N^2 + N)}$. Put $\delta = \min\{\delta_1, \delta_2\}$. $\qquad\square$

We can establish the desire continuity result.

Theorem 7.36. *If $f \in \Im(\mathbf{A})$ and $\varepsilon > 0$, there exists $\delta > 0$ such that if $\mathbf{A}_1 \in L(X)$ with $\|\mathbf{A} - \mathbf{A}_1\| < \delta$, then $f \in \Im(\mathbf{A}_1)$ and $\|f(\mathbf{A}) - f(\mathbf{A}_1)\| < \varepsilon$.*

Proof. Let U be a neighborhood of $\sigma(\mathbf{A})$ where f is analytic and let $V \subset U$ be admissible. By the Perturbation Theorem 7.35 there exists $\delta_1 > 0$ such that $\sigma(\mathbf{A}_1) \subset V$ for $\|\mathbf{A} - \mathbf{A}_1\| < \delta_1$ when $\mathbf{A}_1 \in L(X)$. Thus, $f \in \Im(\mathbf{A}_1)$ when $\|\mathbf{A} - \mathbf{A}_1\| < \delta_1$. Also, by the Perturbation Theorem 7.35, $R_z(\mathbf{A})$ is near $R_z(\mathbf{A}_1)$ for $\|\mathbf{A} - \mathbf{A}_1\|$ small and z not too close to $\sigma(\mathbf{A})$ so we may choose $0 < \delta < \delta_1$, such that

$$\begin{aligned}
\|f(\mathbf{A}) - f(\mathbf{A}_1)\| &= \frac{1}{2\pi} \left\| \int_{\partial V} f(z) \left[R_z(\mathbf{A}) - R_z(\mathbf{A}_1)\right] dz \right\| \\
&\le \frac{1}{2\pi} \int_{\partial V} |f(z)| \|R_z(\mathbf{A}) - R_z(\mathbf{A}_1)\| \, d|z| \\
&< \varepsilon
\end{aligned}$$

for $\|\mathbf{A} - \mathbf{A}_1\| < \delta$. $\qquad\square$

7.6 Abstract Functional Calculus

Although the operational calculus $\theta : \Im(\mathbf{A}) \to L(X)$ described in Theorem 7.33 is an algebra homomorphism which satisfies the continuity property of Theorem 7.32, it does not fit into the formal framework of an abstract functional calculus set forth in Chapter 6 because the space $\Im(\mathbf{A})$ depends on the operator $\mathbf{A}$ and its spectrum $\sigma(\mathbf{A})$. We can, however, use the operational calculus to construct an abstract functional calculus.

Let $\Im$ be the algebra of all entire functions and we equip $\Im$ with the topology of uniform convergence on compact subsets of $\mathbb{C}$. Under this topology $\Im$ is a Fréchet algebra containing the space of polynomials. If $f \in \Im$, then $f \in \Im(\mathbf{A})$ for every $\mathbf{A} \in L(X)$ so $f(\mathbf{A})$ is well-defined. Therefore, we may define an operation calculus

$$\theta : \Im \times L(X) \to L(X)$$

by

$$\theta(f, \ \mathbf{A}) = f(\mathbf{A}).$$

By Theorem 7.27 θ is an algebra homomorphism which satisfies $\theta(p, \ \mathbf{A}) = p(\mathbf{A})$ for every polynomial p and $\mathbf{A} \in L(X)$ so θ is a functional calculus on $L(X)$ relative to $\Im$ (6.1). If $f, g \in \Im$ and $\mathbf{A} \in L(X)$,then

$$\theta(f \circ g, \mathbf{A}) = (f \circ g)(\mathbf{A})$$

and

$$\theta(f, \theta(g, \mathbf{A})) = f(\theta(g, \mathbf{A})) = f(g(\mathbf{A})).$$

By Theorem 7.31,

$$\theta(f \circ g, \mathbf{A}) = \theta(f, \theta(g, \mathbf{A}))$$

so the functional calculus is complete (6.2).

Moreover, by Theorems 7.32 and 7.36, θ is continuous in each variable. It is actually the case that $\theta : \Im \times L(X) \to L(X)$ is jointly continuous, i.e., continuous in both variables.

Theorem 7.37. *The map* $\theta : \Im \times L(X) \to L(X)$ *is jointly continuous.*

Proof. Suppose $f_n \to f$ in $\Im$ and $\mathbf{A}_n \to \mathbf{A}$ in $L(X)$. We have

$$\|f_n(\mathbf{A}_n) - f(\mathbf{A})\| \le \|f_n(\mathbf{A}_n) - f_n(\mathbf{A})\| + \|f_n(\mathbf{A}) - f(\mathbf{A})\|$$

and by Theorem 7.32 the second term on the right hand side of this inequality goes to 0 as $n \to \infty$. Therefore, we need only consider $f_n(\mathbf{A}_n) - f_n(\mathbf{A})$. For this let $\varepsilon > 0$ and $\delta > 0$ be as in Theorem 7.35. Pick N such that $n \ge N$ implies $\|\mathbf{A}_n - \mathbf{A}\| < \delta$. Let $R > \|\mathbf{A}\| + \varepsilon$ and C be the positively oriented circle with center at the origin and radius R. For $n \geq N$ the spectrum of $\mathbf{A}_n$, $\sigma(\mathbf{A}_n)$, lies inside C and

$$\|R_z(\mathbf{A}_n) - R_z(\mathbf{A})\| < \varepsilon$$

for $z \in C$ by Theorem 7.35. Let

$$M = \sup\{|f_n(z)| : z \in C, \ n \geq N\};$$

note $M < \infty$ since $f_n \to f$ uniformly on compact subsets of $\mathbb{C}$. For $n \geq N$,

$$\|f_n(\mathbf{A}_n) - f_n(\mathbf{A})\| = \left\| \frac{1}{2\pi i} \int_C f_n(z) \left[R_z(\mathbf{A}_n) - R_z(\mathbf{A}) \right] dz \right\|$$
$$\leq M R \varepsilon$$

which means $f_n(\mathbf{A}_n) - f_n(\mathbf{A}) \to 0$ as desired. $\qquad \square$

As an example we compute the operational calculus for the Volterra Operator.

Example 7.38. Let $X = C\,[0,\,1]$ and $\mathbf{V} : X \to X$ be the Volterra operator defined by

$$\mathbf{V}x(t) = \int_0^t x(s)ds.$$

Then $\sigma(\mathbf{V}) = \{0\}$,

$$\mathbf{V}^m x(t) = \frac{1}{(m-1)!} \int_0^t (t-s)^{m-1} x(s)ds$$

for $m > 1$ (Cauchy's Formula), and

$$R_z(\mathbf{V}) = \sum_{j=0}^{\infty} \frac{\mathbf{V}^j}{z^{j+1}} \ (z \neq 0).$$

Therefore, if $x = R_z(\mathbf{V})y$, then

$$x(t) = \frac{1}{z}y(t) + \sum_{j=1}^{\infty} \frac{1}{z^{j+1}} \int_0^t \frac{(t-s)^{j-1}y(s)ds}{(j-1)!}$$
$$= \frac{1}{z}y(t) + \frac{1}{z^2} \int_0^t \sum_{j=1}^{\infty} \frac{(t-s)^{j-1}}{z^{j-1}(j-1)!} y(s)ds$$
$$= \frac{1}{z}y(t) + \frac{1}{z^2} \int_0^t e^{(t-s)/z} y(s)ds.$$

If $f \in \Im(\mathbf{V})$, f is analytic on a neighborhood of 0 and if C is a small circle with center at the origin,

$$y(t) = f(\mathbf{V})x(t) = \frac{1}{2\pi i} \int_C f(z) \left[\frac{1}{z}x(t) + \frac{1}{z^2} \int_0^t e^{(t-s)/z} x(s)ds \right] dz$$

$$= f(0)x(t) + \int_0^t x(s) \int_C \frac{1}{2\pi i} f(z) \frac{1}{z^2} e^{(t-s)/z} dz ds.$$

The operational calculus for the Volterra operator can be used to solve certain initial value problems for linear equations. Consider the differential equation.

$$y^{(n)}(s) + a_1 y^{(n-1)}(s) + \ldots + a_n y(s) = x(s)$$

with initial conditions

$$y(0) = y'(0) = \ldots = y^{(n-1)}(0) = 0,$$

for $y \in C[0,1]$ and a_i constants. Then y is a solution to this initial value problem iff

$$(I + a_i \mathbf{V} + \ldots + a_n \mathbf{V}^n)y = \mathbf{V}^n x.$$

If $g(z) = 1 + a_1 z + \ldots + a_n z$ we must solve, $g(\mathbf{V})y = \mathbf{V}^n x$ for y. Since $g(0) \neq 0$, $g(\mathbf{V})$ has an inverse, and we have $y = f(\mathbf{V})x$, where $f(z) = \frac{z^n}{g(z)}$. By the formula above,

$$y(t) = \int_0^t x(s) \left\{ \frac{1}{2\pi i} \int_C \frac{z^{n-2}}{g(z)} e^{\frac{(t-s)}{z}} dz \right\} ds,$$

where the zeros of g are outside C. Changing variables $w = \frac{1}{z}$, this may be written

$$y(t) = \int_0^t x(s) \left\{ \frac{1}{2\pi i} \int_{C'} \frac{e^{(t-s)w}}{w^m + a_1 w^{(m-1)} + \ldots + a_m} dw \right\} ds,$$

where C' is the image of C.

We give two further examples

Example 7.39. Let $X = C[0,1]$ and define $\mathbf{T} : X \to X$ by $\mathbf{T}f(x) = tf(x)$. Then $\sigma(\mathbf{T}) = [0,1]$ and the resolvent operator R_z is given by $R_z f(t) = \frac{1}{z-t} f(t)$ for $z \notin [0,1]$, $f \in X$ [see Example B.22]. Let $f \in \Im(\mathbf{T})$

and C be a simple closed, positively oriented curve with $[0, 1]$ in its interior. Then

$$f(\mathbf{T}) = \frac{1}{2\pi i} \int_C f(z) R_z dz$$

and if $\varphi \in X$ and $t \in [0, 1]$,

$$f(\mathbf{T})\varphi = \frac{1}{2\pi i} \int_C f(z) R_z \varphi dz,$$

$$f(\mathbf{T})\varphi(t) = \frac{1}{2\pi i} \int_C f(z) \frac{1}{z - t} \varphi(t) dt$$

$$= \varphi(t) f(t).$$

Example 7.40. Let $1 \leq p < \infty$ and consider the left shift operator $\mathbf{L}$: $\ell^p \to \ell^p$ defined by $\mathbf{L}\{t_1, t_2, ...\} = \{t_2, t_3, ...\}$, $\{t_j\} \in \ell^p$. Then $\sigma(\mathbf{L}) = \{z : |z| \leq 1\}$ (see Example B.21). The operator $\mathbf{L}$ is represented by the matrix

$$\mathbf{L} = \begin{bmatrix} 0 & 0 & 0 & ... \\ 1 & 0 & 0 & ... \\ 0 & 1 & 0 & ... \\ & \vdots & & \end{bmatrix}$$

and if $|z| > 1$, the resolvent operator $R_z(\mathbf{L})$ is represented by the matrix

$$R_z(\mathbf{L}) = \begin{bmatrix} \frac{1}{z} & 0 & 0 & ... \\ \frac{1}{z^2} & \frac{1}{z} & 0 & ... \\ \frac{1}{z^3} & \frac{1}{z^2} & \frac{1}{z} & ... \\ & \vdots & & \end{bmatrix}.$$

Let $f \in \Im(\mathbf{L})$ and C be a positively oriented circle centered at 0 with radius larger than 1. If $\{t_j\} \in \ell^p$, then

$$f(\mathbf{L})\{t_j\} = \frac{1}{2\pi i} \int_C f(z) R_z(L) \{t_j\} dz$$

$$= \frac{1}{2\pi i} \int_C \left\{ \frac{f(z)}{z} t_1, \frac{f(z)}{z^2} t_1 + \frac{f(z)}{z} t_2, \frac{f(z)}{z^3} t_1 + \frac{f(z)}{z^2} t_2 + \frac{f(z)}{z} t_3, ... \right\} dz$$

$$= \left\{ f(0)t_1, f'(0)t_1 + f(0)t_2, \frac{f''(0)}{2!} f''(0)t_1 + f'(0)t_2 + f(0)t_3, ... \right\}.$$

Thus, the matrix representing $f(L)$ has
$$\left\{ f^k(0)/(k-1)!, f^{(k-1)}(0)/(k-2)!, ..., f'(0), f(0), 0... \right\}$$
in its kth row.

As another example we consider a diagonal operator on ℓ^p.

Example 7.41. Let $\{d_j\}$ be a sequence in C which converges to 0 with $d_j \neq d_k$ for $j \neq k$. Let $\mathbf{D} : \ell^2 \to \ell^2$ be the continuous compact linear operator defined by $\mathbf{D}\{t_j\} = \{d_j t_j\}$, $\{t_j\} \in \ell^2$. $\sigma(\mathbf{D}) = \{d_j : j \in \mathbb{N}\} \cup \{0\}$. If $z \notin \sigma(\mathbf{D})$, then $R_z(\mathbf{D}) = (z - \mathbf{D})^{-1}$ is diagonal operator with $\left\{ \frac{1}{z - d_j} \right\}$ down the diagonal, $R_z(\mathbf{D})\{t_j\} = \left\{ \frac{1}{z - d_j} t_j \right\}$. Let $f \in \Im(\mathbf{D})$, assume f is analytic on an open neighborhood U of $\sigma(\mathbf{D})$ and let V be an admissible neighborhood of U. Put $C = \partial V$. Then

$$f(\mathbf{D})t = \frac{1}{2\pi i} \int_C f(z) R_z(\mathbf{D}) t\, dz = \frac{1}{2\pi i} \int_C \sum_{j=1}^{\infty} \frac{f(z)}{z - d_j} t_j e^j\, dz. \qquad (7.5)$$

Let $M = max\{|f(z)| : z \in C\}$ and $d = dist(C, \sigma(\mathbf{D}))$ so $|z - d_j| \leq d$ for $z \in C$, $j \in \mathbb{N}$. For the jth term in the series on the right hand side of (7.5), we have

$$\left| \frac{f(z)}{z - d_j} \cdot t_j \right|^2 \leq \left| \frac{M}{d} t_j \right|^2$$

so the series converges absolutely and uniformly in ℓ^2.

Therefore, we have
$$f(\mathbf{D})t = \sum_{j=1}^{\infty} \left(\frac{1}{2\pi i} \int_C \frac{f(z)}{z - d_j} dz \right) t_j e^j = \sum_{j=1}^{\infty} f(d_j) t_j e^j.$$

Here,
$$\int_C \frac{f(z)}{z - d_j} dz = 2\pi i f(d_j)$$

since if $C = \bigcup_{k=1}^{\infty} C_k$, where each C_k is a simple closed positively oriented, rectifiable curve, then d_j belongs to the interior of some C_{k_j} so

$$\int_{C_{k_j}} \frac{f(z)}{z - d_j} dz = 2\pi i f(d_j)$$

while
$$\int_{C_k} \frac{f(z)}{z - d_j} dz = 0$$

for $k \neq k_j$. Thus, $f(\mathbf{D})$ is the diagonal operator with $\{f(d_j)\}$ down the diagonal.

7.7 Spectral Sets

A subset $\sigma \subset \sigma(\mathbf{A})$ is *spectral set* if σ is both open and closed in $\sigma(\mathbf{A})$. If σ is a spectral set, then $\sigma(\mathbf{A}) \smallsetminus \sigma = \sigma^c$ is also a spectral set. For example, any isolated point of $\sigma(\mathbf{A})$ is a spectral set.

Suppose σ is a spectral set. Choose $f_\sigma \in \Im(\mathbf{A})$ such that $f_\sigma = 1$ on a neighborhood of σ and $f_\sigma = 0$ on a neighborhood of σ^c. Denote

$$f_\sigma(\mathbf{A}) = \mathbf{P}(\sigma, \ \mathbf{A}) = \mathbf{P}_\sigma(\mathbf{A}) = \mathbf{P}_\sigma$$

and note $\mathbf{P}_\sigma$ is independent of the choice of f_σ. Since $f_\sigma^2 = f_\sigma$ on $\sigma(\mathbf{A})$, $\mathbf{P}_\sigma^2 = \mathbf{P}_\sigma$ (Theorem 7.27) so $\mathbf{P}_\sigma$ is a projection. The function $1 - f_\sigma$ is related to σ^c in the same way as f is related to σ so

$$\mathbf{P}(\sigma^c, \ \mathbf{A}) = \mathbf{P}_{\sigma^c} = \mathbf{P}_{\sigma^c}(\mathbf{A})$$

is a projection and

$$\mathbf{P}_\sigma \mathbf{P}_{\sigma^c} = \mathbf{P}_{\sigma^c} \mathbf{P}_\sigma = 0, \ \mathbf{P}_\sigma + \mathbf{P}_{\sigma^c} = I.$$

Thus, X decomposes as

$$X = \mathbf{P}_\sigma X \oplus \mathbf{P}_{\sigma^c} X = X_1 \oplus X_2.$$

Note also that the operator $\mathbf{A}$ is completely reduced by (X_1, X_2), i.e., $\mathbf{A}X_i \subset X_i$ since $\mathbf{A}\mathbf{P}_\sigma = \mathbf{P}_\sigma \mathbf{A}$, $\mathbf{A}\mathbf{P}_{\sigma^c} = \mathbf{P}_{\sigma^c}\mathbf{A}$.

Example 7.42. Let $\{d_j\}$ be a sequence in $\mathbb{C}$ which converges to 0 with $d_j \neq d_k$ for $j \neq k$, and let $D : \ell^2 \to \ell^2$ be the diagonal operator $\mathbf{D}\{t_j\} = \{d_j t_j\}$ (see Example 7.41). The spectrum of $\mathbf{D}$ is $\sigma(\mathbf{D}) = \{d_j : j \in \mathbb{N}\} \cup \{0\}$ and the resolvent operator is given by

$$R_z(\mathbf{D})\{t_j\} = \left\{(z - d_j)^{-1} t_j\right\}.$$

The projection $\mathbf{P}_{d_j}$ is given by $\mathbf{P}_{d_k}\{t_j\} = t_k e^k$, so we have

$$R_z(\mathbf{D})t = \sum_{j=1}^{\infty} \frac{1}{z - d_j} \mathbf{P}_{d_j} t,$$

$t = \{t_j\} \in \ell^2$. Note this series does not converge in $L(\ell^2)$ but only in the strong operator topology.

Theorem 7.43. *Let* $\mathbf{A}_i = A\mid_{X_i}$. *Then* $\sigma(\mathbf{A}_1) = \sigma$, $\sigma(\mathbf{A}_2) = \sigma^c$,

$$R_z(\mathbf{A})\mid_{X_1} = R_z(\mathbf{A}_1) \text{ and } R_z(\mathbf{A})\mid_{X_2} = R_z(\mathbf{A}_2).$$

The reducing pair (X_1, X_2) *is unique.*

Proof. Let $z \in \sigma$. Suppose $z \notin \sigma(\mathbf{A}_1)$. Then there exists $\mathbf{B} \in L(X_1)$ such that $\mathbf{B}(z - \mathbf{A})x = (z - \mathbf{A})\mathbf{B}x = x$ for $x \in X$. Define $g(t) = 0$ in a neighborhood of σ and $g(t) = \frac{1}{(z-t)}$ in a neighborhood of σ^c. Then

$$g(\mathbf{A})(z - \mathbf{A}) = (z - \mathbf{A})g(\mathbf{A}) = I - \mathbf{P}_\sigma.$$

Extend $\mathbf{A}_1 x = \mathbf{A}\mathbf{P}_\sigma x$ for $x \in X$. Then

$$(z - \mathbf{A})(\mathbf{A}_1 + g(\mathbf{A})) = (\mathbf{A}_1 + g(\mathbf{A}))(z - \mathbf{A}) = I$$

which implies $z \in \rho(\mathbf{A})$, a contradiction. Thus, $\sigma \subset \sigma(\mathbf{A}_1)$.

Suppose $z \notin \sigma_1$. Define $h(t) = \frac{1}{(z-t)}$ in a neighborhood of σ not containing z and $h(t) = 0$ in a neighborhood of σ^c not containing z. Then

$$h(\mathbf{A})(z - \mathbf{A}) = (z - \mathbf{A})h(\mathbf{A}) = \mathbf{P}_\sigma.$$

so

$$h(\mathbf{A})\mid_{X_1} (zI \mid_{X_1} -\mathbf{A}_1) = (zI \mid_{X_1} -\mathbf{A}_1)h(\mathbf{A}) \mid_{X_1} = \mathbf{P}_\sigma \mid_{X_1} = I \mid_{X_1}.$$

This means $z \notin \sigma(\mathbf{A}_1)$ so $\sigma(\mathbf{A}_1) \subset \sigma_1$.

Hence, $\sigma = \sigma(\mathbf{A}_1)$.

Clearly, $R_z(\mathbf{A}_1) = R_z(\mathbf{A}) \mid_{X_1}$.

The other equalities follow.

For uniqueness, suppose (Y_1, Y_2) is another reducing pair for $\mathbf{A}$ satisfying $\sigma(\mathbf{A} \mid_{Y_1}) = \sigma$, $\sigma(\mathbf{A} \mid_{Y_2}) = \sigma^c$. Let C_1 be a Jordan curve with σ inside C_1 and C_2 a Jordan curve with σ^c inside C_2. Then $C = C_1 \cup C_2$ is a Jordan curve with $\sigma(\mathbf{A})$ inside C if C_1 and C_2 are chosen appropriately. If $y \in Y_1$,

$$y = \left[\frac{1}{2\pi i} \int_{C_1} R_z(\mathbf{A} \mid_{Y_1})dz \right] y$$

$$= \left[\frac{1}{2\pi i} \int_{C_1} R_z(\mathbf{A})dz \right] y$$

$$= \left[\frac{1}{2\pi i} \int_{C} f_\sigma(z)R_z(\mathbf{A})dz \right] y$$

$$= f_\sigma(\mathbf{A})y$$

$$= \mathbf{P}_\sigma y.$$

Therefore, $y \in X_1$ so $Y_1 \subset X_1$. Similarly, $Y_2 \subset X_2$.

If $v \in X_1$, then $v = y_1 + y_2$ with $y_i \in Y_i$ so $y_2 = v - y_1 \in X_1 \cap X_2 = \{0\}$. Hence, $v = y_1 \in Y_1$ and $X_1 \subset Y_1$. Therefore, $Y_1 = X_1$. Similarly, we can get $X_2 = Y_2$. $\qquad\square$

For any set σ for which $\mathbf{P}_\sigma$ is defined, let $X_\sigma = \mathbf{P}_\sigma X$ and if $\mathbf{T} \in L(X)$, let $\mathbf{T}_\sigma$ be the restriction of T to X_σ.

Corollary 7.44. *If* $f \in \Im(\mathbf{A})$, *then* $f \in \Im(\mathbf{A}_\sigma)$ *and* $f(\mathbf{A})_\sigma = f(\mathbf{A}_\sigma)$.

Proof. Let U be an admissible neighborhood of the domain of f and $C = \partial U$. By Theorem 7.43,

$$
\begin{aligned}
f(\mathbf{A})_\sigma &= \frac{1}{2\pi i} \left[\int_C f(z) R_z(\mathbf{A}) dz \right]_\sigma \\
&= \frac{1}{2\pi i} \int_C f(z) R_z(\mathbf{A})_\sigma dz \\
&= \frac{1}{2\pi i} \int_C f(z) R_z(\mathbf{A}_\sigma) dz \\
&= f(\mathbf{A}_\sigma).
\end{aligned}
$$

$\square$

Concerning spectral sets for $f(\mathbf{A})$, we have:

Theorem 7.45. *Let* $f \in \Im(\mathbf{A})$ *and* τ *be a spectral set of* $f(\mathbf{A})$. *Then* $\sigma(\mathbf{A}) \cap f^{-1}(\tau)$ *is a spectral set of* $\mathbf{A}$ *and* $\mathbf{P}_\tau(f(\mathbf{A})) = \mathbf{P}_{f^{-1}(\tau)}(\mathbf{A})$.

Proof. Let $f_\tau = 1$ on a neighborhood of τ and $f_\tau = 0$ on a neighborhood of $\tau^c = \sigma(f(\mathbf{A})) \smallsetminus \tau$. Then $f_\tau(f(\mathbf{A})) = \mathbf{P}_\tau(f(\mathbf{A}))$. The Spectral Mapping Theorem implies $\sigma(\mathbf{A}) = f^{-1}(\tau) \cup f^{-1}(\tau^c)$. Since f is continuous, $f^{-1}(\tau)$, $f^{-1}(\tau^c)$ are both open and closed in $\sigma(\mathbf{A})$. Therefore, $\sigma(\mathbf{A}) \cap f^{-1}(\tau)$ is a spectral set of $\mathbf{A}$. If $f_\sigma = f_\tau \circ f$, $\mathbf{P}_\sigma(\mathbf{A}) = f_\sigma(\mathbf{A})$ and Theorem 7.31 gives $\mathbf{P}_\tau(f(\mathbf{A})) = \mathbf{P}_\sigma(\mathbf{A}) = \mathbf{P}_{f^{-1}(\tau)}(\mathbf{A})$. $\square$

7.8 Isolated Points

If z_0 is an isolated point of $\sigma(\mathbf{A})$, then $\{z_0\}$ is a spectral set. Then R_z has a unique Laurent expansion in an annulus $\{z : 0 < |z - z_0| < r\}$ for small r,

$$
R_z(\mathbf{A}) = \sum_{j=-\infty}^{\infty} (z - z_0)^j \mathbf{A}_j,
$$

$\mathbf{A}_j \in L(X)$. We say that z_0 is a *pole of order* k if $\mathbf{A}_{-k} \neq 0$, and $\mathbf{A}_{-j} = 0$ for $j > k$; if $k = 1$, z_0 is called a *simple pole*. If $\mathbf{A}_{-j} \neq 0$ for infinitely many j, then z_0 is an *essential singularity*. For isolated points, we have:

Theorem 7.46. z_0 *is either a pole or an essential singularity.*

Proof. If z_0 were not a singularity, then $\mathbf{A}_{-j} = 0$ for $j \geq 1$ so $\lim_{z \to z_0} R_z(\mathbf{A}) = \mathbf{A}_0 \in L(X)$. But, $(z - \mathbf{A})R_z(\mathbf{A}) = R_z(\mathbf{A})(z - \mathbf{A}) = I$ so letting $z \to z_0$, $(z_0 - \mathbf{A})\mathbf{A}_0 = \mathbf{A}_0(z_0 - \mathbf{A}) = I$ which implies $z_0 \in \rho(\mathbf{A})$. $\square$

We have the following relationships between the coefficients $\mathbf{A}_j$ in the Laurent expansions.

Theorem 7.47.

  (1) $\mathbf{A}_{-1} = I + (\mathbf{A} - z_0)\mathbf{A}_0$
  (2) $\mathbf{A}_{n-1} = (\mathbf{A} - z_0)\mathbf{A}_n, \ n \neq 0$
  (3) $\mathbf{A}_n \mathbf{A} = \mathbf{A}\mathbf{A}_n$
  (4) $(\mathbf{A} - z_0)^n \mathbf{A}_n = \mathbf{A}_0$
  (5) $\mathbf{A}_{-n} = (\mathbf{A} - z_0)^{n-1}\mathbf{A}_{-1}$
  (6) $(\mathbf{A} - z_0)^n \mathbf{A}_{-1} = \mathbf{A}_{-(n+1)}, \ n \geq 1$.

Proof. Multiply $R_z(\mathbf{A}) = \sum\limits_{j=-\infty}^{\infty} (z - z_0)^j \mathbf{A}_j$ on the left by $z - \mathbf{A}$ and use $(z - \mathbf{A})R_z(\mathbf{A}) = I$ to obtain

$$I = \sum_{j=-\infty}^{\infty} (z - \mathbf{A})(z - z_0)^j \mathbf{A}_j$$

$$= \sum_{j=-\infty}^{\infty} [(z - z_0) + (z_0 - \mathbf{A})] (z - z_0)^j \mathbf{A}_j$$

$$= \sum_{j=-\infty}^{\infty} (z - z_0)^{j+1} \mathbf{A}_j + \sum_{j=-\infty}^{\infty} (z_0 - \mathbf{A})(z - z_0)^j \mathbf{A}_j$$

$$= \sum_{j=-\infty}^{\infty} (z - z_0)^j [\mathbf{A}_{j-1} + (z_0 - \mathbf{A})\mathbf{A}_j].$$

The uniqueness of the Laurent expansion gives (1) and (2).
The other equalities follow from the integral formulas for the $\mathbf{A}_j$. $\square$

Definition 7.48. The operator $\mathbf{A}_{-1}$ in the Laurent expansion of $R_z(\mathbf{A})$ about z_0 is called the residue operator [as in classical complex analysis].

We have a characterization of the residue operator as a projection.

Theorem 7.49. $\mathbf{A}_{-1}$ *is the projection operator* $\mathbf{P}_{\{z_0\}} = \mathbf{P}_{z_0}$.

Proof. Choose $r > 0$ such that $\{z : 0 < |z - z_0| < 5r\} \cap \sigma(\mathbf{A}) = \emptyset$.

For $s > 0$ let C_s be the positively oriented circle with center z_0 and radius s and D_s be the disc interior to C_s. Choose $\rho > 5r$ so large that

$$D_r \cup \{z : 2r < |z - z_0| < \rho\} = V \supset \sigma(\mathbf{A})$$

so $\partial V = C_r \cup C_\rho \cup (-C_{2r})$. Then $\mathbf{A}_{-1} = \frac{1}{2\pi i} \int_{C_r} R_z(\mathbf{A}) dz$. Define $f(z) = 0$ if $|z - z_0| > 4r$ and $f(z) = 1$ if $|z - z_0| < 3r$. Then $f \in \Im(\mathbf{A})$ and $f = 1$ on D_r so

$$\begin{aligned}
\mathbf{P}_{z_0} &= f(\mathbf{A}) \\
&= \frac{1}{2\pi i} \int_{\partial V} f(z) R_z(\mathbf{A}) dz \\
&= \frac{1}{2\pi i} \int_{C_r} R_z(\mathbf{A}) dz \\
&= A_{-1}.
\end{aligned}$$ $\qquad\square$

We give examples of poles and essential singularities.

Example 7.50. Consider the Volterra operator $\mathbf{V} : C[0, 1] \to C[0, 1]$, $\mathbf{V}f(t) = \int_0^t f(s) ds$. Then $\sigma(V) = \{0\}$ and

$$R_z f(t) = \frac{1}{z} f(t) + \frac{1}{z^2} \int_0^t e^{(t-s)/z} f(s) ds = \sum_{j=\infty}^{\infty} \frac{\mathbf{V}^j f(t)}{z^{j+1}}, \quad z \neq 0,$$

$$\mathbf{V}^j f(t) = \frac{1}{(j-1)!} \int_0^t (t - s)^{j-1} f(s) ds \ \text{[Example 7.38]}.$$

Hence, $\mathbf{A}_{-1} = \mathbf{P}_{\{0\}}$ and $\mathbf{A}_{-j} = \mathbf{V}^{j-1}$ for $j \geq 2$. This means $\{0\}$ is an essential singularity.

Example 7.51. Let $X = \ell^1$, $\{d_j\} \subset \mathbb{C}$ with $d_j \to 0$ and $d_j \neq d_k$ for $j \neq k$. Let $\mathbf{D} : X \to X$ be the diagonal operator $\mathbf{D}\{t_j\} = \{d_j t_j\}$, $\{t_j\} \in \ell^1$, and the spectrum is $\{d_j : j \in \mathbb{N}\} \cup \{0\}$ so every d_j is an isolated point of the spectrum.

Fix k. We will compute the residue operator with respect to d_k as well as the other operators in the Laurent expansion about d_k. The resolvent operator for $\mathbf{D}$ is $R_z\{t_j\} = \{(z - d_j)^{-1}t_j\}$, $\{t_j\} \in \ell^1$. Let $\delta > 0$ be such that $|d_k - d_j| > \delta$ for $j \neq k$. Restrict z to be in the circle $|z - d_k| < \delta$ so

$$\frac{|z - d_k|}{|d_k - d_j|} < \frac{|z - d_k|}{\delta} < 1. \tag{7.6}$$

Then

$$\frac{1}{z - d_j} = \frac{1}{(d_k - d_j) + (z - d_k)} = \sum_{i=0}^{\infty} \frac{(-1)^i (z - d_k)^i}{(d_k - d_j)^{i+1}}$$

for $k \neq j$, $|z - d_k| < \delta$. Thus,

$$R_z\{t_j\} = \left\{\frac{t_j}{(z - d_j)}\right\} = \sum_{i=0}^{\infty} \frac{1}{z - d_j} t_j e^j$$

$$= \sum_{j=1 \atop j \neq k}^{\infty} \sum_{i=0}^{\infty} \frac{(-1)^i}{(d_k - d_j)^{i+1}} (z - d_k)^i t_j e^j + \frac{1}{z - d_k} t_k e^k. \tag{7.7}$$

This shows that d_k is a simple pole and the residue operator $\mathbf{A}_{-1}$ is given by $\mathbf{A}_{-1}\{t_j\} = t_k e^k = \mathbf{P}_{d_k}\{t_j\}$ as in Theorem 7.49. We next compute the other coefficient operators $\{\mathbf{A}_j : j \geq 0\}$ in the Laurent expansion about d_k. For this we consider the iterated series on the right hand side of equation (7.6). We show this iterated series is absolutely convergent so we may interchange the sums of the series. Consider the absolute values of the "inner series", $\sum_{i=0}^{\infty} \frac{(-1)^i}{(d_k - d_j)^{i+1}} (z - d_k)^i$.

From equation (7.7) we have

$$\sum_{i=0}^{\infty} \left| \frac{(z - d_k)^i}{(d_k - d_j)^{i+1}} \right| = \frac{1}{|d_k - d_j|} \sum_{i=0}^{\infty} \left| \frac{z - d_k}{d_k - d_j} \right|^i$$

$$= \frac{1}{|d_k - d_j| - |z - d_k|} \leq \frac{1}{\delta - |z - d_k|}$$

and since $\{t_j\} \in \ell^1$, the iterated series in equation (7.7) is absolutely convergent. Interchanging the order of summation in equation (7.7) gives

$$R_z\{t_j\} = \sum_{i=0}^{\infty} \left\{ \sum_{j=1 \atop j \neq k}^{\infty} \frac{(-1)^i}{(d_k - d_j)^{i+1}} t_j e^j \right\} (z - d_k)^i + \frac{1}{z - d_k} t_k e^k.$$

Therefore, the ith coefficient operator, $\mathbf{A}_i$, in the Laurent expansion is given

by $\mathbf{A}_i \{t_j\} = \sum_{\substack{j=1 \\ j \neq k}}^{\infty} \frac{(-1)^i}{(d_k - d_j)^{i+1}} t_j e^j = \left\{ \sum_{\substack{j=1 \\ j \neq k}}^{\infty} \frac{(-1)^i}{(d_k - d_j)^{i+1}} t_j \right\}_{j=1}^{\infty}$.

That is, $\mathbf{A}_i$ is the diagonal operator defined by $\{a_j^i\}_j$, where $a_j^i = \frac{(-1)^i}{(d_k - d_j)^{i+1}}$ if $j \neq k$ and $a_j^i = 0$. Note that the series defining $\mathbf{A}_i$ converges in the strong operator topology but not in norm.

We next give a characterization of the poles of the resolvent. For this we need the notion of ascent and descent of an operator.

Definition 7.52. The ascent, $\alpha(\mathbf{A})$, of the operator $\mathbf{A}$ is the smallest integer k such that $\ker \mathbf{A}^k = \ker \mathbf{A}^{k+1}$. If there exists no such integer, we set $\alpha(\mathbf{A}) = \infty$. The descent of the operator $\mathbf{A}$, denoted by $\delta(\mathbf{A})$, is the smallest integer, $\delta(\mathbf{A})$, such that $\mathcal{R}\mathbf{A}^k = \mathcal{R}\mathbf{A}^{k+1}$. If there exists no such integer, we set $\delta(\mathbf{A}) = \infty$.

We have the following important property of an operator with finite ascent and descent.

Theorem 7.53. *If the operator $\mathbf{A}$ has finite ascent and descent, then*

$$p = \alpha(\mathbf{A}) = \delta(\mathbf{A}).$$

Moreover,

$$X = \ker \mathbf{A}^p \oplus \mathcal{R}\mathbf{A}^p,$$

$((\ker \mathbf{A}^p)$, $\mathcal{R}\mathbf{A}^p)$ *reduce $\mathbf{A}$, and $\mathbf{A}$ is nilpotent on $\ker \mathbf{A}^p$ and invertible on $\mathcal{R}\mathbf{A}^p$.*

Proof. Let $k = \alpha(\mathbf{A}) < \infty$ and $\ell = \delta(\mathbf{A}) < \infty$. First note

$$\ker \mathbf{A}^p \cap \mathcal{R}\mathbf{A}^k = \{0\} \quad \text{for} \quad p \in \mathbb{N} \tag{7.8}$$

and

$$X = \ker \mathbf{A}^\ell + \mathcal{R}\mathbf{A}^q \text{ for } q \in \mathbb{N}. \tag{7.9}$$

For equation (7.8), fix p and let $y \in \ker \mathbf{A}^p \cap \mathcal{R}\mathbf{A}^k$. Then there exists $x \in X$ such that $y = \mathbf{A}^k x$. This implies $\mathbf{A}^{k+p} x = \mathbf{A}^p y = 0$ or $x \in \ker \mathbf{A}^{k+p} = \ker \mathbf{A}^k$. Hence, $y = \mathbf{A}^k x = 0$.

For equation (7.9) fix q and let $x \in X$. Then $\mathbf{A}^\ell x \in \mathcal{R}\mathbf{A}^\ell = \mathcal{R}\mathbf{A}^{\ell+q}$ so there exists $z \in X$ such that $\mathbf{A}^\ell x = \mathbf{A}^{\ell+q}z$. Then $\mathbf{A}^\ell(x - \mathbf{A}^q z) = 0$ so $x - \mathbf{A}^q z \in \ker \mathbf{A}^\ell$. Note $x = (x - \mathbf{A}^q z) + \mathbf{A}^q z$.

Setting $p = \ell$ in equation (7.8) and $q = k$ in equation (7.9) gives

$$X = \ker \mathbf{A}^\ell \oplus \mathcal{R}\mathbf{A}^q \text{ for } q \in \mathbb{N}. \tag{7.10}$$

Fix $u \in \mathbf{A}^{l+1}$. By equation (7.10), $u = u_1 + u_2$, where $u_1 \in \ker \mathbf{A}^\ell \subset \ker \mathbf{A}^{\ell+1}$ and $u_2 \in \mathcal{R}\mathbf{A}^k$. From equation (7.8) we have $u_2 = u - u_1 \in \ker \mathbf{A}^{k+1} \cap \mathcal{R}\mathbf{A}^k = \{0\}$ so $u_2 = 0$. Hence $u = u_1 \in \ker \mathbf{A}^\ell$ so $\ker \mathbf{A}^{\ell+1} = \ker \mathbf{A}^\ell$. This means $k \leq \ell$.

For the other inequality, let $u \in \mathcal{R}\mathbf{A}^k$. From equation (7.9), $u = u_1 + u_2$ with $u_1 \in \ker \mathbf{A}^\ell$ and $u_2 \in \mathcal{R}\mathbf{A}^{k+1} \subset \mathcal{R}\mathbf{A}^k$. From equation (7.8) $u_1 = u - u_2 \in \mathcal{R}\mathbf{A}^k \cap \ker \mathbf{A}^\ell = \{0\}$ so $u_1 = 0$ and $u = u_2 \in \mathcal{R}\mathbf{A}^{k+1}$. Hence, $\mathcal{R}\mathbf{A}^k = \mathcal{R}\mathbf{A}^{k+1}$ so $\ell \leq k$, and we have the equality $k = \ell$. $\qquad\square$

It is actually the case that $\mathcal{R}\mathbf{A}^p$ in the theorem is closed so the decomposition in the theorem involves closed subspaces and the associated projections are continuous. To show this we need two lemmas.

Let Y be a Banach space and $\mathbf{T} \in L(X, Y)$. Let $\|\cdot\|_q$ be the quotient norm on $X/\ker \mathbf{T}$ and $[x] = [x + \ker \mathbf{T}]$ denote cosets. We define a norm $\|\cdot\|_\mathbf{T}$ on $\mathcal{R}\mathbf{T}$ by setting

$$\|y\|_\mathbf{T} = \|\mathbf{T}x\|_\mathbf{T} = \|y\| + \|[x]\|_q,$$

where $y = \mathbf{T}x \in \mathcal{R}\mathbf{T}$.

Lemma 7.54.

(1) $(\mathcal{R}\mathbf{T}, \|\cdot\|_\mathbf{T})$ is a Banach space.

(2) $\mathcal{R}\mathbf{T}$ is closed iff the norms $\|\cdot\|$ and $\|\cdot\|_\mathbf{T}$ are equivalent on $\mathcal{R}\mathbf{T}$.

Proof.   (1). Suppose $\{y_n\}$ is $\|\cdot\|_\mathbf{T}$ Cauchy in $\mathcal{R}\mathbf{T}$ with $y_n = \mathbf{T}x_n$, $x_n \in X$. Now

$$\|y_n - y_m\|_\mathbf{T} = \|y_n - y_m\| + \|[x_n - x_m]\|_q$$

implies $\{y_n\}$ is $\|\cdot\|$ Cauchy and $\{[x_n]\}$ is $\|\cdot\|_q$ Cauchy. Therefore, there exist $y \in Y, x \in X$ such that $\|y_n - y\| \to 0$, $\|[x_n - x]\|_q \to 0$. There exist $z_n \in X$ such that $\|z_n\| \to 0$ and $\mathbf{T}z_n = \mathbf{T}x_n - \mathbf{T}x = y_n - \mathbf{T}x$. Since $\|y_n - y\| \to 0$ and $\|\mathbf{T}z_n\| \to 0$, $y = \mathbf{T}x \in \mathcal{R}\mathbf{T}$ and

$$\|y_n - y\|_\mathbf{T} = \|y_n - y\| + \|[x_n - x]\|_q \to 0.$$

Hence, $(\mathcal{R}\mathbf{T}, \; \|\cdot\|_{\mathbf{T}})$ is complete.

(2). The identity $I : (\mathcal{R}\mathbf{T}, \|\cdot\|_{T}) \to (\mathcal{R}\mathbf{T}, \|\cdot\|)$ is continuous so if $\mathcal{R}\mathbf{T}$ is closed (complete), the Open Mapping Theorem implies $\|\cdot\|_{\mathbf{T}}$ and $\|\cdot\|$ are equivalent.

Conversely, if $\|\cdot\|_{\mathbf{T}}$ and $\|\cdot\|$ are equivalent, by (1) $(\mathcal{R}\mathbf{T}, \; \|\cdot\|)$ is complete and closed. $\qquad\qquad\square$

Lemma 7.55. T *has closed range iff there exists a closed subspace* $Z \subset Y$ *such that* $\mathcal{R}\mathbf{T} \cap Z = \{0\}$ *and* $\mathcal{R}\mathbf{T} \oplus Z$ *is closed.*

Proof. If $\mathcal{R}\mathbf{T}$ is closed, $Z = \{0\}$ will satisfy the condition.

Conversely, assume $Z \subset Y$ is a closed subspace such that $\mathcal{R}\mathbf{T} \cap Z = \{0\}$ and $V = \mathcal{R}\mathbf{T} \oplus Z$ is closed. The map $\mathbf{S} : \mathcal{R}\mathbf{T} \to V/Z$ defined by $\mathbf{S}y = [y] = y + Z$ is linear, one-one and onto. Moreover, $\mathbf{S} : (\mathcal{R}\mathbf{T}, \; \|\cdot\|_{\mathbf{T}}) \to \left(V/Z, \; \|\cdot\|_q\right)$, where $\|\cdot\|_q$ is the quotient norm, is continuous since $\|\mathbf{S}y\|_q \leq \|y\| \leq \|y\|_{\mathbf{T}}$ when $y = \mathbf{T}x \in \mathcal{R}\mathbf{T}$. Since both of these spaces are complete (Lemma 7.54), $\mathbf{S}$ has a continuous inverse so there exists $c > 0$ such that $c \|y\|_{\mathbf{T}} \leq \|\mathbf{S}y\|$ for $y \in \mathcal{R}\mathbf{T}$. Thus, Lemma 7.54 gives the result. $\qquad\square$

If now follows from Lemma 7.55 that the subspace $\mathcal{R}\mathbf{A}^p$ in Theorem 7.53 is closed.

We can now give a characterization of poles.

Theorem 7.56. *The following are equivalent:*

(1) z_0 *is a pole or order* k,
(2) $(\mathbf{A} - z_0)^{k-1} \mathbf{A}_{-1} \neq 0$ *and* $(\mathbf{A} - z_0)^k \mathbf{A}_{-1} = 0$,
(3) *The ascent and descent of the operator* $z_0 - \mathbf{A}$ *are both finite and equal to the order of the pole* z_0.

Proof. (1) $\implies$ (2) : We have $\mathbf{A}_{-k} \neq 0$ and $\mathbf{A}_{-m} = 0$ for $m > k$. From Theorem 7.47, $(\mathbf{A} - z_0)^{k-1} \mathbf{A}_{-1} = \mathbf{A}_{-k} \neq 0$ and $(\mathbf{A} - z_0)^k \mathbf{A}_{-1} = \mathbf{A}_{-(k+1)} = 0$.

(2) $\implies$ (3) : From Theorem 7.49, $\mathbf{A}_{-1} = \mathbf{P}_{z_0}$ so $(\mathbf{A} - z_0)^{k-1} \mathbf{P}_{z_0} \neq 0$ and $(\mathbf{A} - z_0)^k \mathbf{P}_{z_0} = 0$.

We claim $\ker (z_0 - \mathbf{A})^k = \ker (z_0 - \mathbf{A})^{k+1}$. We have
$$\ker (z_0 - \mathbf{A})^k \subset \ker (z_0 - \mathbf{A})^{k+1}.$$

For the reverse inclusion, let $x \in \ker (z_0 - \mathbf{A})^{k+1}$ and suppose $x \notin \ker (z_0 - \mathbf{A})^k$. Then $y = (z_0 - \mathbf{A})^k x \neq 0$ and
$$(z_0 - \mathbf{A}) y = (z_0 - \mathbf{A})^{k+1} x = 0$$

or $\mathbf{A}y = z_0 y$ and $f(\mathbf{A})y = f(z_0)\,y$ for any $f \in \Im(\mathbf{A})$. In particular, for $f = f_{\{z_0\}}$ in the definition of $\mathbf{P}_{z_0}$, $\mathbf{P}_{z_0}y = f_{\{z_0\}}y = y$ which gives

$$0 = (\mathbf{A} - z_0)^k \,\mathbf{P}_{z_0}x = \mathbf{P}_{z_0}\,(\mathbf{A} - z_0)^k\,x = \mathbf{P}_{z_0}y = y \neq 0,$$

a contradiction. This establishes the claim.

This shows $(z_0 - \mathbf{A})$ has finite ascent and $\alpha = \alpha\,(z_0 - \mathbf{A}) \leq k$. In particular, $\ker\,(z_0 - \mathbf{A})^{\alpha} = \ker\,(z_0 - \mathbf{A})^n$ for $n \geq \alpha$.

Since $(\mathbf{A} - z_0)^{k-1}\,\mathbf{P}_{z_0} \neq 0$, there exists an x in the range of $\mathbf{P}_{z_0}$ such that $(\mathbf{A} - z_0)^{k-1}\,x = (\mathbf{A} - z_0)^{k-1}\,\mathbf{P}_{z_0}x \neq 0$. Since

$$(\mathbf{A} - z_0)^k\,x = (\mathbf{A} - z_0)^{k-1}\,\mathbf{P}_{z_0}x = 0,$$

$\ker\,(z_0 - \mathbf{A})^{k-1} \neq \ker\,(z_0 - \mathbf{A})^k$. Thus, $\alpha \geq k$ and $\alpha = k$.

Next, consider the decomposition in Theorem 7.43 for the spectral set $\{z_0\}$ and the operator $\mathbf{A}$, $X = X_1 \oplus X_2$, where $X_1 = \mathbf{P}_{z_0}X$, $X_2 = \mathbf{P}_{\{z_0\}^c}X$, $\sigma\,(\mathbf{A}\,|_{X_1}) = \{z_0\}$, $\sigma\,(\mathbf{A}\,|_{X_2}) = \sigma\,(\mathbf{A}) \setminus \{z_0\}$. This means, the operator $z_0 - \mathbf{A}$ is invertible on X_2 and then also $(z_0 - \mathbf{A})^n$ is invertible on X_2.

Since $(\mathbf{A} - z_0)^k\,\mathbf{P}_{z_0} = 0$, $(z_0 - \mathbf{A})^k = 0$ on X_1. Hence,

$$\mathcal{R}\,(z_0 - \mathbf{A})^{k+1} = \mathcal{R}\,(z_0 - \mathbf{A})^k = X_2$$

so $(z_0 - \mathbf{A})$ has finite descent, and we must have $\alpha\,(z_0 - \mathbf{A}) = \delta\,(z_0 - \mathbf{A}) = k$.

(3) $\implies$ (1) : Assume $k = \alpha\,(z_0 - \mathbf{A}) = \delta\,(z_0 - \mathbf{A}) < \infty$. If $Y = \ker\,(z_0 - \mathbf{A})^k$, $Z = \mathcal{R}\,(z_0 - \mathbf{A})^k$, then Y and Z are $\mathbf{A}$ invariant, closed subspaces such that $X = Y \oplus Z$ and $z_0 - \mathbf{A}$ is quasinilpotent on Y and invertible on Z (Theorem 7.53).

Since $\mathbf{A} = \mathbf{A}\,|_Y \oplus \mathbf{A}\,|_Z$, $\sigma(\mathbf{A}) = \sigma\,(\mathbf{A}\,|_Y) \cup \sigma\,(\mathbf{A}\,|_Z)$. Since $z_0 \notin \sigma\,(\mathbf{A}\,|_Z)$ and $\sigma\,(\mathbf{A}\,|_Y) = \{z_0\}$, $\sigma\,(\mathbf{A}\,|_Z) = \sigma\,(\mathbf{A}) \setminus \{z_0\}$. By the uniqueness in Theorem 7.43, $Y = X_1$, $Z = X_2$. In particular, $\mathbf{P}_{z_0}$ is the projection of X on Y along Z. If $n > k$ and $x \in X$, then since

$$\mathbf{P}_{z_0}x \in Y = \ker\,(z_0 - \mathbf{A})^k = \ker\,(z_0 - \mathbf{A})^{k-1},$$

we have $\mathbf{A}_{-n}x = (\mathbf{A} - z_0)^{n-1}\,\mathbf{P}_{z_0}x = 0$. Thus, $\mathbf{A}_{-n} = 0$ for $n > k$ and z_0 is a pole of order less than or equal to k. But, from (2) $\implies$ (3), we must have the order of the pole is k. $\qquad\square$

We have several corollaries of Theorem 7.56.

If $\mathbf{A}$ is a compact operator, then every non-zero point of the spectrum, $\sigma(\mathbf{A})$, is isolated, and we have:

Corollary 7.57. *If $\mathbf{A}$ is compact, then every non-zero point of $\sigma(\mathbf{A})$ is a pole.*

Proof. For any $0 \neq z \in \mathbb{C}$, the operator $z - \mathbf{A}$ has finite ascent and descent ([TL] V.7.9) so Theorem 7.56 applies. $\qquad \square$

Corollary 7.58. *If z_0 is a pole of order k, then z_0 is an eigenvalue of A. Moreover,*

$$\mathcal{R}\mathbf{A}_{-1} = \ker\left(z_0 - \mathbf{A}\right)^n, \ \mathcal{R}\left(I - \mathbf{A}_{-1}\right) = \mathcal{R}\left(z_0 - \mathbf{A}\right)^n$$

for $n \geq k$ so in particular, $X = \ker\left(z_0 - \mathbf{A}\right)^n \oplus \mathcal{R}\left(z_0 - \mathbf{A}\right)^n$ for $n \geq k$.

Proof. By Theorem 7.56, $(\mathbf{A} - z_0)^{k-1}\mathbf{A}_{-1} \neq 0$ and $(\mathbf{A} - z_0)^k \mathbf{A}_{-1} = 0$. Pick $x \in X$ such that $y = (\mathbf{A} - z_0)^{k-1}\mathbf{A}_{-1}x \neq 0$. Then

$$(\mathbf{A} - z_0)\, y = (\mathbf{A} - z_0)^k \mathbf{A}_{-1}x = 0$$

so $\mathbf{A}y = z_0 y$ and z_0 is an eigenvalue with eigenvector y.

Set $X_1 = \ker \mathbf{A}_{-1} = \mathcal{R}\left(I - \mathbf{A}_1\right)$, $X_2 = \mathcal{R}\mathbf{A}_{-1}$ and $N_n = \ker\left(z_0 - \mathbf{A}\right)^n$, $\mathcal{R}_n = \mathcal{R}\left(z_0 - \mathbf{A}\right)^n$.

First we claim $X_2 = N_n$. For this let $x \in N_n$. From Theorem 7.47,

$$(\mathbf{A} - z_0)^n \mathbf{A}_{n-1} = (\mathbf{A} - z_0)\mathbf{A}_0 = \mathbf{A}_{-1} - I$$

so

$$0 = \mathbf{A}_{n-1}\left(\mathbf{A} - z\right)^n x = (\mathbf{A} - z_0)^n \mathbf{A}_{n-1}x = (\mathbf{A} - z_0)\mathbf{A}_0 x = \mathbf{A}_{-1}x - x$$

which implies $x = \mathbf{A}_{-1}x \in X_2$. Hence, $N_n \subset X_2$. Next, let $x \in X_2$. Then $x = \mathbf{A}_{-1}y$ for some y and

$$x = \mathbf{A}_{-1}y = \mathbf{A}_{-1}\mathbf{A}_{-1}y = \mathbf{A}_{-1}x.$$

By Theorem 7.47, $(\mathbf{A} - z_0)^n \mathbf{A}_{-1} = \mathbf{A}_{-(n+1)}$ so $(\mathbf{A} - z_0)^n x = \mathbf{A}\mathbf{A}_{-(n+1)}x$. But, $\mathbf{A}_{-(n+1)} = 0$ for $n \geq k$ so $(\mathbf{A} - z_0)^n x = 0$ for $n \geq k$. Therefore, $X_2 \subset N_n$ for $n \geq k$ and we have $X_2 = N_n$ for $n \geq k$.

By Theorem 7.47, $(\mathbf{A} - z_0)^n \mathbf{A}_{n-1} = \mathbf{A}_{-1} - I$ which implies $X_1 \subset \mathcal{R}_n$. Next, observe that if $x \in \mathcal{R}_n \cap N_n$, $n \geq k$, then $x = 0$. For if $x = (\mathbf{A} - z_0)^n y$ and $(\mathbf{A} - z_0)^n x = 0$ by the equality above $y \in N_{2n} = N_n = X_2$ and $x = 0$. Suppose $x \in \mathcal{R}_n$. Then $x = x_1 + x_2$ where $x_1 = (I - \mathbf{A}_{-1})x \in X_1, x_2 = \mathbf{A}_{-1}x$ since $X = X_1 \oplus X_2$. Now $X_1 \subset \mathcal{R}_n$ by the first part so $x_2 = x - x_1 \in \mathcal{R}_n$. But, $x_2 \in X_2 = N_n$ by the equality above so $x_2 \in \mathcal{R}_n \cap N_n$ and $x_2 = 0$ by the observation above. Therefore $x = x_1 \in X_1$ and $\mathcal{R}_n \subset X_1$. Hence, $X_1 = \mathcal{R}_n$ for $n \geq k$. $\qquad \square$

For any set σ for which $\mathbf{P}_\sigma$ is defined let $X_\sigma = \mathbf{P}_\sigma X$ (so $\mathbf{A}X_\sigma \subset X_\sigma$) and $\mathbf{A}_\sigma = \mathbf{A}\mid_{X_\sigma}$.

Corollary 7.59. *Let σ be a spectral set of $\sigma(\mathbf{A})$. A point $z \in \sigma$ is a pole of order k iff z is a pole of $\mathbf{A}_\sigma$ of order k.*

Proof. By Theorem 7.56, z is a pole of order k iff $(z - \mathbf{A})^{k-1} \mathbf{P}_z \neq 0$ and $(z - \mathbf{A})^k \mathbf{P}_z = 0$. Since $z \in \sigma$, $\mathbf{P}_z \mathbf{P}_\sigma = \mathbf{P}_z$ and thus $(zI - \mathbf{A})^j \mathbf{P}_z = (zI_\sigma - \mathbf{A}_\sigma)^j \mathbf{P}_z$. This gives the result. $\square$

Corollary 7.60. *If X is finite dimensional, then every point of the spectrum of $\mathbf{A}$ is a pole.*

Proof. For any point $z \in \mathbb{C}$, $z - \mathbf{A}$ has finite ascent and descent. Since every point of the spectrum is isolated, Theorem 7.56 applies. $\square$

We use the Riesz operational calculus to characterize those functions for which $f(\mathbf{A}) = 0$.

Theorem 7.61. (Minimal Equation) *Let $f \in \Im(\mathbf{A})$. Then $f(\mathbf{A}) = 0$ if and only if $f = 0$ on an open set containing $\sigma(\mathbf{A})$ except possibly for a finite number of poles of $R_z(\mathbf{A}) = R_z$ at the points $z_1, ..., z_k$ and if the poles exist and the order of the pole at z_j is m_j, then f has a zero of order at least m_j at z_j.*

Proof. $\Longleftarrow$: From the definition of the operational calculus,

$$f(\mathbf{A}) = \frac{1}{2\pi i} \sum_{j=1}^{k} \int_{C_j} f(z) R_z dz,$$

where C_j is a small circle centered at z_j. If f has a zero of order at least m_j at z_j, then $f(z) R_z$ is analytic inside C_j and $\int_{C_j} f(z) R_z dz = 0$ by the Cauchy integral formula. Hence, $f(\mathbf{A}) = 0$.

$\Longrightarrow$: Suppose $f(\mathbf{A}) = 0$. By the Spectral Mapping Theorem,

$$\{0\} = f(\sigma(\mathbf{A})) = \sigma(f(\mathbf{A})).$$

Let f be analytic on the open neighborhood U of $\sigma(\mathbf{A})$. For each $z \in \sigma(\mathbf{A})$, there exists $\varepsilon_z > 0$ such that the spheres

$$S(z, \varepsilon_z) = \{w : |w - z| < \varepsilon_z\} \subset U,$$

and since $\sigma(\mathbf{A})$ is compact, a finite number of the spheres, say, $S(z_1, \varepsilon_{z_1}), ..., S(z_n, \varepsilon_{z_k})$, cover $\sigma(\mathbf{A})$. If some $S(z_j, \varepsilon_{z_j})$ contains an infinite number of points of $\sigma(\mathbf{A})$, then $f \equiv 0$ on $S(z_j, \varepsilon_{z_j})$ since f is analytic.

Thus, if U_1 is the union of the spheres containing infinitely many points of $\sigma(\mathbf{A})$, $f \equiv 0$ on U_1. Hence, U_1 contains all but a finite number of isolated points of $\sigma(\mathbf{A})$, say, $z_1, ..., z_k$.

Suppose f is not identically zero on a neighborhood of z_1. Since $f(\sigma(\mathbf{A})) = \{0\}$, f has a zero of finite order n at z_1. Therefore,

$$g_1(z) = \frac{(z_1 - z)^n}{f(z)}$$

is analytic on a neighborhood of z_1. Let $h \in \Im(\mathbf{A})$ be 1 on a neighborhood of z_1, and 0 on a neighborhood of $\sigma(\mathbf{A}) \setminus \{z_1\}$ and set $g = g_1 h$. Then

$$f(\mathbf{A})g(\mathbf{A}) = (z_1 - \mathbf{A}) \, h(\mathbf{A}).$$

The Laurent expansion of R_z in a neighborhood $0 < |z_1 - z| < \varepsilon$ is

$$R_z = \sum_{k=-\infty}^{\infty} \mathbf{A}_k \, (z - z_j)^k \,,$$

$$\mathbf{A}_k = -\frac{1}{2\pi i} \int_{C_1} (z_1 - z)^{-(k+1)} R_z dz = - (z_1 - \mathbf{A})^{-(k+1)} \, h(\mathbf{A}),$$

where C_1 is a small circle centered at z_1. Hence,

$$\mathbf{A}_{-(k+1)} = - (z_1 - \mathbf{A})^k \, h(\mathbf{A}) = 0$$

for $k \geq n$, and z_1 is a pole of order less than or equal to n.

Similar statements hold for the other z_j so f is either identically 0 on a neighborhood of each z_j or z_j is a pole of order m_j and f has a zero of order at least m_j at z_j. $\qquad\square$

Corollary 7.62. *Let* $z_1, \ldots, z_k$ *be poles of* $\mathbf{A}$ *with orders* $n_1, \ldots, n_k$, *respectively. Set* $\sigma = \{z_1, \ldots, z_k\}$. *If* $f \in \Im(\mathbf{A})$, *then*

$$f(\mathbf{A})\mathbf{P}_\sigma = \sum_{j=1}^{k} \sum_{\ell=1}^{n_j-1} \frac{f^{(\ell)}(z_j)}{\ell!} (\mathbf{A} - z_j)^\ell \, \mathbf{P}_{z_j}.$$

Proof. Set $g(z) = \sum_{j=1}^{k} \sum_{\ell=1}^{n_j-1} \frac{f^{(\ell)}(z_j)}{\ell!} (z - z_j)^\ell$. Then $g^{(\ell)}(z_j) = f^{(\ell)}(z_j)$ for $0 \leq \ell \leq n_j - 1$, $1 \leq j \leq k$, so the equality $f(\mathbf{A})\mathbf{P}_\sigma = g(\mathbf{A})\mathbf{P}_\sigma$ follows from the Minimal Equation. $\qquad\square$

As an application of the Riesz operational calculus we prove Wiener's Theorem for functions with absolutely convergent Fourier series.

7.9   Wiener's Theorem

Let $\mathcal{W}$ be the set of all functions $f : \mathbb{R} \to \mathbb{C}$ which are periodic, with period 2π, and have absolutely convergent Fourier series. That is, $f(t) = \sum\limits_{k\in\mathbb{Z}} a_k e^{ikt}$, where $a_k = \int\limits_{-\pi}^{\pi} f(s)e^{-iks}ds$ and $\sum\limits_{k\in\mathbb{Z}} |a_k| < \infty$. We define a norm on $\mathcal{W}$ by

$$\|f\| = \sum_{k\in\mathbb{Z}} |a_k|.$$

$\mathcal{W}$ is clearly a vector space and is actually an algebra under pointwise multiplication. Indeed, if $f(t) = \sum\limits_{k\in\mathbb{Z}} a_k e^{ikt}$, $g(t) = \sum\limits_{k\in\mathbb{Z}} b_k e^{ikt}$, where $f, g \in \mathcal{W}$, then

$$f(t)g(t) = \sum_{k\in\mathbb{Z}} \sum_{j\in\mathbb{Z}} a_j b_{k-j} e^{ikt}$$

so

$$\sum_{k\in\mathbb{Z}} \left| \sum_{j\in\mathbb{Z}} a_j b_{k-j} \right| \le \sum_{k\in\mathbb{Z}} \sum_{j\in\mathbb{Z}} |a_j|\, |b_{k-j}| = \sum_{j\in\mathbb{Z}} |a_j| \sum_{k\in\mathbb{Z}} |b_j| = \|f\|\,\|g\|.$$

This gives $\|fg\| \le \|f\|\,\|g\|$ and $\mathcal{W}$ is a normed algebra under this product. Actually, $\mathcal{W}$ is complete so $\mathcal{W}$ is a Banach algebra.

If $f \in \mathcal{W}$, then f induces a linear operator $\mathbf{T}_f : \mathcal{W} \to \mathcal{W}$ defined by $\mathbf{T}_f g = fg$. Since $\|\mathbf{T}_f g\| = \|fg\| \le \|f\|\,\|g\|$, $\mathbf{T}_f$ is continuous and $\|\mathbf{T}_f\| \le \|f\|$. But, $g \equiv 1 \in \mathcal{W}$ and $\|g\| = 1$, $\|\mathbf{T}_f g\| = \|f\|$ so $\|\mathbf{T}_f\| = \|f\|$. We also have

$$\left\| \mathbf{T}_f^n \right\| = \left\| \mathbf{T}_{f^n} \right\| = \|f^n\|$$

so

$$r(\mathbf{T}_f) = \lim_n \left\| \mathbf{T}_f^n \right\|^{\frac{1}{n}} = \lim_n \|f^n\|^{\frac{1}{n}},$$

where $r(\mathbf{T}_f)$ is the spectral radius of $\mathbf{T}_f$. We now claim

$$r(\mathbf{T}_f) = M_f = \sup\{|f(t)| : t \in \mathbb{R}\}.$$

Since

$$|f(t)| = \left| \sum_{k\in\mathbb{Z}} a_k e^{ikt} \right| \le \sum_{k\in\mathbb{Z}} |a_k| = \|f\|,$$

we have $M_f \le \|f\|$. Also, $M_f^n = M_{f^n}$ so $M_f = M_{f^n}^{\frac{1}{n}} \le \|f^n\|^{\frac{1}{n}}$. Hence, $M_f \le r(\mathbf{T}_f)$.

The reverse inequality is more difficult. For this assume first that f is a trigonometric polynomial, $f(t) = \sum_{k=1}^{P} a_{j_k} e^{ij_k t}$. In the product $f(t)^n = \sum_{j=1}^{N} d_{n_j} e^{in_j t}$, the number N of terms with distinct exponents n_j will be less than or equal to $\binom{p+n-1}{p-1}$. Note $\|f^n\| = \sum_{j=1}^{N} |d_{n_j}| \le \left(\sum_{j=1}^{N} |d_{n_j}|^2 \right)^{\frac{1}{2}} N^{\frac{1}{2}}$. From Bessel's Equality,

$$\int_0^{2\pi} |f(t)|^{2n}\, dt = \int_0^{2\pi} f(t)^n \overline{f(t)^n} dt = \sum_{j=0}^{N} |d_{n_j}|^2 \le M_f^{2n} \cdot 2\pi.$$

Therefore,

$$\|f^n\|^{\frac{1}{n}} \le \left(\sum_{j=1}^{N} |d_{n_j}|^2 \right)^{\frac{1}{2n}} N^{\frac{1}{2n}} \le M_f \, (2\pi N)^{\frac{1}{2n}}.$$

Letting $n \to \infty$ gives

$$r(\mathbf{T}_f) = \lim_n \|f^n\|^{\frac{1}{n}} \le M_f$$

so the desired inequality is established for trigonometric polynomials.

For the general case, if $f \in \mathcal{W}$ and $\varepsilon > 0$, there exists a trigonometric polynomial p such that $\|f - p\| < \varepsilon$. Set $h = f - p$ so $M_h \le \|h\| < \varepsilon$ and $r(\mathbf{T}_p) = M_p \le M_f + M_h$. Since $r(\mathbf{T}_h) = \lim \|h^n\|^{\frac{1}{n}} \le \|h\|$,

$$r(\mathbf{T}_f) \le r(\mathbf{T}_p) + r(\mathbf{T}_h) \le M_p + \|h\| < M_f + 2\epsilon$$

and $r(\mathbf{T}_f) \le M_f$. Hence, $r(\mathbf{T}_f) = M_f$ for every $f \in \mathcal{W}$.

The equality $M_f = r(\mathbf{T}_f)$ is due to Beurling ([Beu]). We thus have that $\sigma(\mathbf{T}_f)$ is contained in the disc $\{z : |z| \le M_f\}$. More precisely, we have:

Theorem 7.63. $\sigma(\mathbf{T}_f)$ *is the range of f,* $\{f(t) : t \in \mathbb{R}\} = \mathcal{R}f.$

Proof. First, suppose $z \in \rho(\mathbf{T}_f)$. Then for every $h \in \mathcal{W}$, there exists $g \in \mathcal{W}$ such that $(z - \mathbf{T}_f) g = h$. In particular, for $h = 1$, there exists $g \in \mathcal{W}$ such that $(z - \mathbf{T}_f) g = 1$ or $(z - f) g = 1$. Hence, $z - f(t) \ne 0$ for $t \in \mathbb{R}$ so $z \notin \mathcal{R}f$. Therefore, $\mathcal{R}f \subset \sigma(\mathbf{T}_f)$.

For the other containment, suppose $z \notin \mathcal{R}f$. Set

$$\delta = dist\,(z,\ \mathcal{R}f) = \inf\,\{|z - f(t)| : t \in \mathbb{R}\};$$

$\delta > 0$ since $\mathcal{R}f$ is compact. Also, set $\Delta = \sup\{|z - f(t)| : t \in \mathbb{R}\}$; note $\Delta < \infty$. Consider the function

$$f^*(t) = |f(t) - z|^2 - \frac{1}{2}\left(\Delta^2 + \delta^2\right);$$

$f^* \in \mathcal{W}$ since $(f(\cdot) - z)\overline{(f(\cdot) - z)} \in \mathcal{W}$. Now

$$\sup\{|f^*(t)| : t \in \mathbb{R}\} = \Delta^2 - \frac{1}{2}\left(\Delta^2 + \delta^2\right) = \frac{1}{2}\left(\Delta^2 - \delta^2\right)$$

so $r(\mathbf{T}_{f^*}) = M_{f^*} = \frac{1}{2}(\Delta^2 - \delta^2)$. If $w = -\frac{1}{2}(\Delta^2 + \delta^2)$, then $|w| = \frac{1}{2}(\Delta^2 + \delta^2) > r(\mathbf{T}_{f^*})$ so $w \in \rho(\mathbf{T}_{f^*})$. Therefore,

$$\mathbf{T}_{f^*} - w = (\mathbf{T}_{f^*-w}) = (\mathbf{T}_{\overline{f}} - \overline{z})(\mathbf{T}_f - z)$$

is invertible. Hence $z \in \rho(\mathbf{T}_f)$ and $\sigma(\mathbf{T}_f) \subset \mathcal{R}f$. $\qquad\square$

We can now obtain the theorem of Wiener.

Theorem 7.64. (Wiener) *If $f \in \mathcal{W}$ is such that $f(t) \neq 0$ for every $t \in \mathbb{R}$, then $\frac{1}{f} \in \mathcal{W}$.*

Proof.   By hypothesis, $0 \notin \mathcal{R}f = \sigma(\mathbf{T}_f)$ so $\mathbf{T}_f$ is invertible. In particular, there exists $g \in \mathcal{W}$ such that $\mathbf{T}_f g = 1 = fg$ so $\frac{1}{f} = g \in \mathcal{W}$. $\qquad\square$

There is a very beautiful proof of Wiener's Theorem using Banach algebras due to Gelfand. See [Sw2] 43.12 or [TL] VII.5.3.

As another application of the Riesz operational calculus for the operators $\mathbf{T}_f$ we derive a theorem of Levi. By Theorem 7.63 an analytic function u belongs to $\mathcal{F}(\mathbf{T}_f)$ if u is analytic on a neighborhood of the range of f, $\mathcal{R}f$. It follows from the proof of Theorem 7.63 that the resolvent operator of $\mathbf{T}_f$ is given by

$$R_z(\mathbf{T}_f)h = (z - f)^{-1}h = \mathbf{T}_{1/(z-f)}h \text{ for } h \in \mathcal{W}, \ z \in \rho(\mathbf{T}_f).$$

Let u in $\mathcal{F}(\mathbf{T}_f)$ and let V be admissible for the domain of u with C the boundary of V. If $h \in \mathcal{W}$,

$$u(\mathbf{T}_f)h = \frac{1}{2\pi i}\int_C u(z)R_z(\mathbf{T}_f)h\,dz = \frac{1}{2\pi i}\int_C u(z)(z - f)^{-1}h\,dz,$$

since $C \subset \rho\left(\mathbf{T}_f\right)$. Applying the Dirac delta (Example E.3) function, δ_t, to this equation gives

$$u\left(\mathbf{T}_f\right) h\left(t\right) = \frac{1}{2\pi i} \int_C u\left(z\right)\left(z - f\left(t\right)\right)^{-1} h\left(t\right) dz = u\left(f\left(t\right)\right) h\left(t\right)$$

by Cauchy's Formula. That is,

$$u\left(\mathbf{T}_f\right) h = \left(u \circ f\right) h$$

or

$$u\left(\mathbf{T}_f\right) = \mathbf{T}_{u \circ f}.$$

Thus, we have:

Theorem 7.65 (Levi). *If $f \in \mathcal{W}$ and u is analytic on a neighborhood of $\mathcal{R}\mathbf{T}_f = \sigma\left(\mathbf{T}_f\right)$, then the composition $u \circ f$ belongs to $\mathcal{W}$.*

See [Wie] for Levi's theorem.

We give a final application of the Riesz operational calculus to compact operators.

Theorem 7.66. *Let f be a non-constant entire function with $f(z) \neq 0$ for $z \neq 0$. Suppose $f(\mathbf{A})$ is compact and $0 \neq z_0 \in \sigma(\mathbf{A})$. Then z_0 is an isolated point of $\sigma(\mathbf{A})$ and is a pole. Moreover, $\mathbf{P}_{z_0} X$ is finite dimensional and if k is the order of the pole, $X = \ker(z_0 - \mathbf{A})^k \oplus \mathcal{R}(z_0 - \mathbf{A})^k$.*

Proof. Suppose $\{z_k\} \subset \sigma(\mathbf{A})$, $z_k \neq z_0$, and $z_k \to z_0$. By the Spectral Mapping Theorem 7.30, $f(z_k) \in \sigma(f(\mathbf{A}))$, $f(z_k) \to f(z_0)$ and since f is non-constant, $f(z_k) \neq f(z_0)$ for large k. Since $f(\mathbf{A})$ is compact, $f(z_0) = 0$ and $z_0 = 0$ by hypothesis. This means any non-zero point of $\sigma(A)$ is isolated.

Let C be a small circle with center at z_0. Let $g \in \mathcal{F}(\mathbf{A})$ be equal to $1/f(z)$ for z inside C and $g(z) = 0$ for z outside C. Then $fg = f_{\{z_0\}}$ so $f(\mathbf{A})g(\mathbf{A}) = \mathbf{P}_{z_0}$, and since $f(\mathbf{A})$ is compact, $\mathbf{P}_{z_0}$ is a compact projection and has finite dimensional range. Let $\mathbf{A}_{z_0}$ be the restriction of $\mathbf{A}$ to $\mathbf{P}_{z_0} X = X_{z_0}$. Since $\{z_0\} = \sigma(\mathbf{A}_{z_0})$ (Theorem 7.43) and $\mathbf{A}_{z_0}$ is compact, z_0 is a pole of $\mathbf{A}_{z_0}$ (Corollary 7.57) and is, therefore, a pole of $\mathbf{A}$ (Corollary 7.59). Corollary 7.58 gives the final conclusion. $\qquad\square$

Remark 7.67. Note that the decomposition in the theorem above gives a Fredholm Alternative for the operator $z_0 - \mathbf{A}$, i.e., $z_0 - \mathbf{A}$ is one-to-one iff $z_0 - \mathbf{A}$ is onto. The theorem in particular applies to the function $f(z) = z^n, n \in \mathbb{N}$, so if the operator $\mathbf{A}^n$ is compact for some n, then the theorem and the Fredholm Alternative applies. For example, any weakly compact operator $\mathbf{T}$ from $L^1[0,1]$ into itself or any weakly compact operator $\mathbf{T}$ from $C(S)$ into itself has a compact square $\mathbf{T}^2$ (see [DS] VI.8.13, VI.7.5 and the remarks and references on page 541).

Chapter 8

Weyl's Functional Calculus

Around 1928 Hermann Weyl in [H. Weyl, The theory of groups and quantum mechanics, Dover N.Y. 1931] defined a functional calculus to deal with the unbounded self adjoint operators of differentiation and multiplication by a position coordinate. Usually the phase space is $\mathbb{R}^n \times \mathbb{R}^n$ and states are represented by elements $(p, q) \in \mathbb{R}^n \times \mathbb{R}^n$ with $p = (p_1, \ldots, p_n)$ the momentum vector, and $q = (q_1, \ldots, q_n)$ the position vector. A classical observable is represented by a real function defined on the phase space. Using von Neumann's spectral theorem for unbounded self adjoint operators one can identify quantum observables with self adjoint, in general unbounded, operators on some Hilbert space $\mathcal{H}$, usually $L^2(\mathbb{R}^n)$. In this view the spatial coordinate projection $(p, q) \longrightarrow q_j$ is mapped to the multiplication operator $\mathbf{X}_j$ of multiplication by x_j. The momentum projection $(p, q) \longrightarrow p_j$ corresponds to the differential operator $\mathbf{D}_j = \frac{h}{2\pi i}\partial_{x_j}$. Weyl proposed that given $(p, q) \in \mathbb{R}^n \times \mathbb{R}^n$, the exponential function $e^{-2\pi i(qx + p\xi)}$ should be assigned to the operator $e^{-2\pi i(q\mathbf{X} + p\mathbf{D})}$ which exists as a unitary operator on $L^2(\mathbb{R}^n)$. Then given a suitable function f defined on the phase space, and its Fourier transform $\hat{f}$ we can consider

$$f(x, \xi) = \int_{\mathbb{R}^{2n}} e^{2\pi i(qx + p\xi)} \hat{f}(p, q) \, dp \, dq$$

from which we can define $f(\mathbf{X}, \mathbf{D})$ formally as the Bochner integral

$$f(\mathbf{X}, \mathbf{D}) = \int_{\mathbb{R}^{2n}} e^{2\pi i(q\mathbf{X} + p\mathbf{D})} \hat{f}(p, q) \, dp \, dq \qquad (8.1)$$

which makes sense as an evaluation of the function $f(p, q)$ on the pair of operators $\mathbf{X}$, $\mathbf{D}$. In particular, if f is in the Schwartz space $\mathcal{S}(\mathbb{R}^{2n})$

(Definition E.19), then its Fourier transform $\widehat{f}$ is Borel-measurable and integrable so (8.1) can make sense as an operator-valued distribution.

This motivates the generalization to n-tuples of operators. Let f be a "suitable" function in $\mathbb{R}^n$ and $(\mathbf{A}_1, \ldots, \mathbf{A}_n)$ an n-tuple of operators in the Banach algebra $L(X)$ where X is a Hilbert space and the $\mathbf{A}_i$ are self adjoint. The map $f(x_1, \ldots, x_n) = f(\mathbf{A}_1, \ldots, \mathbf{A}_n)$ is defined as follows:

$$f(\mathbf{A}_1, \ldots, \mathbf{A}_n) = \int_{\mathbb{R}^n} e^{2\pi i \xi \mathbf{A}} \widehat{f}(\xi) d\xi \, , \tag{8.2}$$

where $\xi \mathbf{A} = (\xi_1 \mathbf{A}_1 + \xi_2 \mathbf{A}_2 + \cdots + \xi_n \mathbf{A}_n)$.

The exponential in (8.2) makes sense as a unitary operator on X (Theorem 4.12) and we may again interpret (8.2) as an operator valued tempered distribution with values in $L(X)$. The Banach valued distribution are defined in the same way as in the real case (Appendix E).

A critical result concerning the Weyl functional calculus is that it gives consistent results with the natural algebraic definition of the polynomial. However, the functional calculus given by (8.2) is not multiplicative on the variable f. That is to say, the product $f_1 \cdot f_2$ is not mapped, in general, to $f_1(\mathbf{A}) \cdot f_2(\mathbf{A})$.

The Weyl calculus for bounded self adjoint operators is considered in [Ne] and [Tay]. The key idea of the argument is the application of the Paley–Wiener Theorem E.52 to growth estimates for exponentials of operators. The main references for this chapter are [An], [AZ], [Ey] and [Tay].

Now we would like to get a transform that will be more general than the transforms defined previously but that agrees with the previous functional calculi defined for polynomials. That is we would like the extension, W, to satisfy $W(p, \mathbf{A}) = p(\mathbf{A})$. The most natural space to consider for the Weyl transform is the space

$$F = \{f \in L^1(\mathbb{R}) : \widehat{f} \in L^1(\mathbb{R})\}$$

and if $f \in F$ and $\mathbf{A}$ is a bounded self adjoint operator, the Weyl transform is defined by

$$W(f, \mathbf{A}) = \int_{\mathbb{R}} e^{2\pi i t \mathbf{A}} \widehat{f}(t) dt.$$

The integral makes sense as a Bochner integral since $e^{2\pi i t \mathbf{A}}$ is a unitary operator of norm 1 and $\widehat{f} \in L^1(\mathbb{R})$. We define a norm of F by $\|f\| =$

$\left\|\widehat{f}\right\|_1$. The Weyl transform defined on this space has the following continuity properties.

Proposition 8.1. *Let $f \in F$. If $\{\mathbf{A}_k\}$ is a sequence of self adjoint operators which is norm convergent to the self adjoint operator $\mathbf{A}$, then $W(f, \mathbf{A}_k) \to W(f, \mathbf{A})$ in norm.*

Proof. For each t, $\exp(2\pi ti\mathbf{A}_k) \to \exp(2\pi ti\mathbf{A})$ in norm by Theorem 7.36 so the proposition follows from the Dominated Convergence Theorem 5.20. □

Similarly, we have:

Proposition 8.2. *Let $f \in F$. If $\{\mathbf{A}_k\}$ is sequence of self adjoint operators which is convergent in the strong operator topology to the self adjoint operator $\mathbf{A}$, then $W(f, \mathbf{A}_k) \to W(f, \mathbf{A})$ in the strong operator topology.*

Proposition 8.3. *Let $\mathbf{A}$ be self adjoint. The linear map $f \to W(f, \mathbf{A})$ from F into $L(X)$ is norm continuous.*

Proof. We have $\|W(f, \mathbf{A})\| \le \|f\|$ so the result is immediate. □

To have more insight on this functional calculus we will start by studying it for a single bounded self adjoint operator $\mathbf{A}$ in $L(X)$ where X is a Hilbert space. For a function $f \in \mathcal{S}(\mathbb{R})$ we define

$$W(f, \mathbf{A}) = \int_{-\infty}^{\infty} e^{2\pi it\mathbf{A}} \widehat{f}(t)dt \qquad (8.3)$$

which is well defined as a Bochner integral in $L(X)$ since $e^{2\pi it\mathbf{A}}$ is unitary and continuous and $\widehat{f}$ is continuous, $\widehat{f} \in \mathcal{S}(\mathbb{R})$ (Theorem E.35). Furthermore,

$$\|W(f, \mathbf{A})\| \le \int_{-\infty}^{\infty} \|e^{2\pi it\mathbf{A}} \widehat{f}(t)\|dt \le \int_{-\infty}^{\infty} |\widehat{f}(t)|dt = \|\widehat{f}\|_1 \ .$$

At this point there is nothing that will justify us to write the integral on (8.3) as $f(\mathbf{A})$. We will provide some type of justification. We'd like that the class of function for which $f(\mathbf{A})$ is well defined contains at least the class of polynomials $\mathscr{P}(\mathbb{R})$ and be consistent with the definition of $p(\mathbf{A})$, given in previous chapters for $p \in \mathscr{P}(\mathbb{R})$. Perhaps the shortest way to accomplish this is using the spectral representation of the operator $\mathbf{A}$ (Theorem 2.7).

Take $f \in \mathcal{S}(\mathbb{R})$, its Fourier transform

$$\widehat{f}(\xi) = \int_{\mathbb{R}} f(\xi) e^{-2\pi i \xi x} dx$$

belongs also to $\mathcal{S}(\mathbb{R})$. Take $\mathbf{A} \in L(X)$ to be self adjoint. By Theorem 4.12 the operator valued function $\xi \longrightarrow e^{2\pi i \xi \mathbf{A}}$ maps ξ to a unitary operator, for every $\xi \in \mathbb{R}$. Let us consider the definition:

$$W(f, \mathbf{A}) = \int_{\mathbb{R}} \widehat{f}(\xi) e^{2\pi i \xi \mathbf{A}} d\xi . \tag{8.4}$$

This definition makes sense as a Bochner integral of the operator valued function $\widehat{f}(\xi) e^{2\pi i \xi \mathbf{A}}$ which is strongly measurable since $\widehat{f}$ and $e^{2\pi i \xi \mathbf{A}}$ are both continuous. Now, since $\|\widehat{f}(\xi) e^{2\pi i \xi \mathbf{A}}\| = |\widehat{f}(\xi)|$, we have that $W(f, \mathbf{A})$ is well defined.

Now, the function $x \longrightarrow e^{2\pi i \xi x}$ is continuous, hence measurable and bounded on an interval containing $\sigma(\mathbf{A})$. Then, if E denotes the spectral measure for the operator $\mathbf{A}$, by Theorem 2.7 we have:

$$e^{2\pi i \xi \mathbf{A}} = \int_{\mathbb{R}} e^{2\pi i \xi t} dE(t) , \tag{8.5}$$

where the integral is convergent in the sense of the norm.
If we substitute (8.5) in (8.4) we get:

$$W(f, \mathbf{A}) = \int_{\mathbb{R}} \int_{\mathbb{R}} \widehat{f}(\xi) e^{2\pi i \xi t} dE(t) d\xi .$$

Note that $|\widehat{f}(\xi) e^{2\pi i \xi t}| = |\widehat{f}(\xi)|$ for all $(\xi, t) \in \mathbb{R} \times \mathbb{R}$. Now, since $\widehat{f}(\xi) \in L^1(\mathbb{R})$ and the projection valued measure $E(t)$ is finite, we can apply Theorem 5.24 and reverse the order of integration:

$$W(f, \mathbf{A}) = \int_{\mathbb{R}} \int_{\mathbb{R}} \widehat{f}(\xi) e^{2\pi i \xi t} d\xi dE(t)$$
$$= \int_{\mathbb{R}} f(t) dE(t) = f(\mathbf{A})$$

and since $f \in \mathcal{S}(\mathbb{R})$ clearly $\int_{\mathbb{R}} f(t) dE(t)$ is well defined and the two definitions coincide $W(f, \mathbf{A}) = f(\mathbf{A})$.

Now, we can extend the class of functions f to those $f \in L^1(\mathbb{R})$ such that $\widehat{f} \in L^1(\mathbb{R})$. The main question now is to see what happens if f is a polynomial p since $\widehat{p}$ may only make sense as a tempered distribution. The natural map between $\mathbf{A}$ and $p(\mathbf{A})$ will be recovered if we consider

$e^{2\pi i \xi \mathbf{A}}$ as an operator valued tempered distribution. This interpretation of the multivariable case of n-self adjoint operators holds even if they do not commute.

Now, we will prove that $W(p, \mathbf{A})$ for p a polynomial coincides with $p(\mathbf{A})$ as defined in Chapter 7. This will be made in a sequence of lemmas.

Lemma 8.4. *Let $p(\xi)$ be a polynomial and $\varphi \in \mathcal{S}(\mathbb{R})$ be a function such that $\varphi(0) = 1$. Then*

$$\lim_{j \to \infty} W\left(\left(p(\xi) \cdot \varphi\left(\frac{\xi}{j}\right)\right), \mathbf{A}\right) = p(\mathbf{A})$$

in $L(X)$.

Proof. It is sufficient to prove the lemma for

$$p(\xi) = \xi^m, \quad m = 0, 1, \ldots .$$

We will also denote by $\mathcal{F}$ the Fourier transform. If $m > 0$ we need to calculate $W\left(\xi^m \cdot \varphi\left(\frac{\xi}{j}\right), \mathbf{A}\right)$.
For this we will use the fact that

$$\mathcal{F}\left[\xi^m \varphi\left(\frac{\xi}{j}\right)\right](t) = \frac{1}{(2\pi i)^m}\left(\frac{d}{dt}\right)^m j\widehat{\varphi}(jt)$$

which is obtained by repeated differentiation under the integral sign (Theorem E.33 (c)). Then

$$W\left(\xi^m \cdot \varphi\left(\frac{\xi}{j}\right), \mathbf{A}\right) = \int_{-\infty}^{\infty} e^{2\pi it\mathbf{A}} \mathcal{F}\left[\xi^m \varphi\left(\frac{\xi}{j}\right)\right] dt$$

$$= \int_{-\infty}^{\infty} e^{2\pi it\mathbf{A}} \frac{1}{(2\pi i)^m} j\left(\frac{d}{dt}\right)^m \widehat{\varphi}(jt) \, dt$$

$$= \int_{-\infty}^{\infty} \mathbf{A}^m e^{2\pi it\mathbf{A}} j\widehat{\varphi}(jt) \, dt .$$

The last equality is obtained by integration by parts.
So we finally have

$$W\left(\xi^m \cdot \varphi\left(\frac{\xi}{j}\right), \mathbf{A}\right) = \mathbf{A}^m \int_{-\infty}^{\infty} e^{2\pi it\mathbf{A}} j\widehat{\varphi}(jt) \, dt .$$

We want to prove that

$$\lim_{j \to \infty} \left\|\left(W\left(\xi^m \cdot \varphi\left(\frac{\xi}{j}\right), \mathbf{A}\right) - \mathbf{A}^m\right)\right\| = 0 . \tag{8.6}$$

Now

$$W\left(\xi^m \cdot \varphi\left(\frac{\xi}{j}\right), \mathbf{A}\right) - \mathbf{A}^m = \mathbf{A}^m \left[\int_{-\infty}^{\infty} e^{2\pi it\mathbf{A}} j\widehat{\varphi}(jt)\, dt - \mathbf{I}\right].$$

If we prove that the bracket tends to zero if $j \to \infty$ the limit (8.6) will be zero.

Now

$$\int_{-\infty}^{\infty} j\widehat{\varphi}(jt)\, dt = \int_{-\infty}^{\infty} j\frac{1}{j}\widehat{\varphi}(t)\, dt = \int_{-\infty}^{\infty} e^{i0}\widehat{\varphi}(t)\, dt = \varphi(0) = 1\,,$$

where we used the inverse formula. We have:

$$\int_{-\infty}^{\infty} e^{2\pi it\mathbf{A}} j\varphi(jt)\, dt - \mathbf{I} = \int_{-\infty}^{\infty} e^{2\pi it\mathbf{A}} j\widehat{\varphi}(jt)\, dt - \int_{-\infty}^{\infty} j\widehat{\varphi}(jt)\, dt\mathbf{I}$$

$$= \int_{-\infty}^{\infty} \left(e^{2\pi it\mathbf{A}} - \mathbf{I}\right) j\widehat{\varphi}(jt)\, dt\,.$$

To calculate this integral we divide it in two. Take $M > 0$, M will be fixed later. We need to evaluate

$$\int_{-\infty}^{\infty} \left(e^{2\pi it\mathbf{A}} - \mathbf{I}\right) j\widehat{\varphi}(jt)\, dt =$$

$$\int_{|t|\leq M} \left(e^{2\pi it\mathbf{A}} - \mathbf{I}\right) j\widehat{\varphi}(jt)\, dt + \int_{|t|>M} \left(e^{2\pi it\mathbf{A}} - \mathbf{I}\right) j\widehat{\varphi}(jt)\, dt$$

and prove that each integral converges to 0, for a suitable M when $j \to 0$. For the first integral we have:

$$e^{ax} - 1 = 1 + ax + \frac{(ax)^2}{2!} + \cdots - 1 = ax + \frac{(ax)^2}{2!} + \frac{(ax)^3}{3!} + \cdots$$

$$= ax\left(1 + \frac{ax}{2!} + \frac{(ax)^2}{3!} + \cdots\right) \leq axe^{ax}$$

so

$$\left\|e^{2\pi it\mathbf{A}} - \mathbf{I}\right\| \leq 2\pi|t|\|\mathbf{A}\|e^{2\pi|t|\|\mathbf{A}\|}$$

since $|t| \leq M$

$$\leq 2\pi M\|\mathbf{A}\|e^{2\pi M\|\mathbf{A}\|}$$

and for $M \to 0$ the above approaches to 0. Then

$$\left\|\int_{|t|\leq M} \left(e^{2\pi it\mathbf{A}} - \mathbf{I}\right) j\widehat{\varphi}(jt)\, dt\right\| \longrightarrow 0 \quad \text{if} \quad M \to 0.$$

Now for the second term $\int_{|t|>M} \left(e^{-2\pi it\mathbf{A}} - \mathbf{I}\right) j\widehat{\varphi}(jt)\, dt$ we have:

$$\left\|e^{2\pi it\mathbf{A}} - \mathbf{I}\right\| \leq \left\|e^{2\pi it\mathbf{A}}\right\| + \|\mathbf{I}\| \leq 2$$

since $e^{2\pi it\mathbf{A}}$ is unitary (Theorem 4.12). Then:

$$\left\| \int_{t>M} \left(e^{2\pi it\mathbf{A}} - \mathbf{I} \right) j\widehat{\varphi}(jt)\, dt \right\| \le 2 \int_{t>\frac{M}{j}} |\widehat{\varphi}(t)|\, dt$$

and this last term tends to zero if M is fixed and j tends to infinity. It remains to prove the claim for $m = 0$. This can be done using the same type of argument as before so

$$\lim_{j\to\infty} W\left(\varphi\left(\tfrac{\xi}{j}\right), \mathbf{A} \right) = \mathbf{I}$$

and

$$\lim_{j\to\infty} W\left(\left(p(\xi) \cdot \varphi\left(\tfrac{\xi}{j}\right) \right), \mathbf{A} \right) = p(\mathbf{A}) .$$

$\square$

Lemma 8.5. *There exists $M > 0$, in fact it can be taken as $\|\mathbf{A}\|$, such that if $\psi \in C_0^\infty(\mathbb{R})$, the space of smooth functions with compact support, and $\mathrm{supp}\,(\psi) \subset \{\, \xi \in \mathbb{R} : |\xi| > M \,\}$, then $W(\psi, \mathbf{A}) = 0$.*

Proof.   Given $\varphi \in \mathcal{S}(\mathbb{R})$, such that $\varphi(0) = 1$ and $\psi \in C_0^\infty$ we have that the sequence $\left\{\, \varphi\left(\tfrac{t}{j}\right) \widehat{\psi}(t) : j \in \mathbb{N} \,\right\}$ is pointwise convergent to $\widehat{\psi}(t)$, since φ is bounded and $\widehat{\psi} \in \mathcal{S}(\mathbb{R}) \subset L^1(\mathbb{R})$, the Dominated Convergence Theorem implies:

$$\lim_{j\to\infty} \int_{-\infty}^{\infty} e^{2\pi it\mathbf{A}} \varphi\left(\tfrac{t}{j}\right) \widehat{\psi}(t)\, dt = \int_{-\infty}^{\infty} e^{2\pi it\mathbf{A}} \widehat{\psi}(t) = W(\psi, \mathbf{A}) .$$

Now for each $j \in \mathbb{N}$:

$$\int_{-\infty}^{\infty} e^{2\pi it\mathbf{A}} \varphi\left(\tfrac{t}{j}\right) \widehat{\psi}(t)\, dt = \int_{-\infty}^{\infty} \varphi\left(\tfrac{t}{j}\right) e^{2\pi it\mathbf{A}} \left(\int_{-\infty}^{\infty} e^{-2\pi its\mathbf{I}} \psi(s)\, ds \right) dt$$

$$= \int_{-\infty}^{\infty} \varphi\left(\tfrac{t}{j}\right) \left(\int_{-\infty}^{\infty} e^{2\pi it(\mathbf{A}-s\mathbf{I})} \psi(s)\, ds \right) dt .$$

By Fubini's Theorem

$$\int_{-\infty}^{\infty} e^{2\pi it\mathbf{A}} \varphi\left(\tfrac{t}{j}\right) \widehat{\psi}(t)\, dt = \int_{-\infty}^{\infty} \left(\int_{-\infty}^{\infty} e^{2\pi it(\mathbf{A}-s\mathbf{I})} \varphi\left(\tfrac{t}{j}\right) dt \right) \psi(s)\, ds$$

$$= \int_{-\infty}^{\infty} u_j(s)\psi(s)\, ds ,$$

where

$$u_j(s) = \int_{-\infty}^{\infty} e^{2\pi i t(\mathbf{A} - s\mathbf{I})} \varphi \left(\frac{t}{j} \right) dt \ .$$

We will reach our goal if we can find $M > 0$ independent of j such that $u_j(s) = 0$ for $|s| > M$. We will do that by choosing properly the function φ.

Let $\rho \in \mathcal{C}_0^{\infty}$ be such that $0 \leq \rho \leq 1$, $supp(\rho) \subset \{s : |s| \leq 1\}$ and $\int_{-\infty}^{\infty} \rho(s) \, ds = 1$.
We take $\varphi = \widehat{\rho}$ that is:

$$\varphi(t) = \widehat{\rho}(t) = \int_{-\infty}^{\infty} e^{-2\pi i x t} \rho(x) \, dx \quad \text{and} \quad \varphi\left(\frac{t}{j}\right) = \int_{-\infty}^{\infty} e^{-2\pi i t x} j \rho(j x) \, dx \ .$$

Clearly $\varphi(0) = 1$.

Claim 8.6. φ *can be extended to an analytic function in* $\mathbb{C}$.

Let $\varphi(z) = \widehat{\rho}(z) = \int_{-\infty}^{\infty} e^{-2\pi i x z} \rho(x) dx$ for $z \in \mathbb{C}$.
This makes sense because:

$$\left| e^{-2\pi i x z} \right| = \left| e^{-2\pi i x \, \mathrm{Re}(z)} e^{2\pi x \, \mathrm{Im}(z)} \right| \leq \left| e^{2\pi x \, \mathrm{Im}(z)} \right| \leq e^{2\pi |\mathrm{Im}(z)|},$$

since $|x| < 1$. So the integral converges absolutely. We can take any derivative we like and bring it under the integral sign, thanks to the bound we just proved and the compact support of ρ.

Claim 8.7. *For every* N *there is a constant* $C_N > 0$ *such that* $|\varphi(z)| \leq C_N e^{|\mathrm{Im}(z)|} (1 + |z|)^{-N}$.

We have seen that $\widehat{\rho}$ is analytic in $\mathbb{C}$, and because of this it clearly satisfies a bound in a given compact set, like the unit disk. So we only need to prove the inequality for $|z| > 1$.
We can write:

$$\widehat{\rho}(z) = \int_{-\infty}^{\infty} e^{-2\pi i x z} \rho(x) \, dx$$

$$= \int_{-\infty}^{\infty} \frac{1}{(-2\pi i z)^k} \frac{d^k}{dx^k} \left(e^{-2\pi i x z} \right) \rho(x) \, dx$$

integrating by parts $\qquad = \int_{-\infty}^{\infty} \frac{1}{(-2\pi i z)^k} e^{-2\pi i x z} \frac{d^k}{dx^k} \rho(x) \, dx \ .$

Now $\rho \in C_0^\infty$ so for each $k \in \mathbb{N}$, so there exists C_k such that $\left|\frac{d^k}{dx^k}\rho(x)\right| < C_k$. Then

$$
\begin{aligned}
|(2\pi i z)^k \widehat{\rho}(z)| &\leq \left| \int_{-\infty}^{\infty} e^{-2\pi i x z} \frac{d^k}{dx^k}\rho(x)\, dx \right| \\
&\leq \sup_{|x| \leq 1} \left| e^{-2\pi i x z} \right| \int_{|x| \leq 1} \left| \frac{d^k}{dx^k}\rho(x) \right|\, dx \\
&\leq e^{2\pi |\mathrm{Im}(z)|} 2C_k
\end{aligned}
$$

so

$$
|2\pi z|^k |\varphi(z)| \leq e^{2\pi |\mathrm{Im}(z)|} 2C_k \,.
$$

Taking $N = k$, since $|z| > 1$, we have

$$
|\varphi(z)| \leq 2C_N e^{2\pi |\mathrm{Im}(z)|}(1 + |z|)^{-N} \tag{8.7}
$$

and from that for every $j \in \mathbb{N}$ we have

$$
\left| \varphi\left(\frac{z}{j}\right) \right| \leq C_N j^N (1+|z|)^{-N} e^{2\frac{\pi |\mathrm{Im}(z)|}{j}} \,. \tag{8.8}
$$

Claim 8.8. $u_j(s) = \int_{-\infty}^{\infty} e^{2\pi i t (\mathbf{A} - s\mathbf{I})} \varphi\left(\frac{t}{j}\right)\, dt = 0$ *if* $|s| \geq \|\mathbf{A}\|$.

To evaluate this improper integral we will use complex integration. The function $z \longrightarrow e^{2\pi i z (\mathbf{A} - s\mathbf{I})} \varphi\left(\frac{z}{j}\right)$ is analytic. Then it's integral along the rectangle with vertices $-R$, R, $R + i\eta$, $-R + i\eta$ with $\eta > 0$ and $R > 0$ is zero [Cauchy's Theorem]. We will prove that the integrals on the vertical segments tend to zero when $R \to \infty$, and then u_j can also be represented by the integral on the complex line $t \to t + i\eta$ since it is independent of η. Both integrals are evaluated in the same way.

Note that since the exponents commute:

$$
e^{2\pi i (R + i y)(\mathbf{A} - s\mathbf{I})} = e^{2\pi i R(\mathbf{A} - s\mathbf{I})} e^{-2\pi y(\mathbf{A} - s\mathbf{I})} \,. \tag{8.9}
$$

Now

$$
\left\| \int_R^{R + i\eta} e^{2\pi i z (\mathbf{A} - s\mathbf{I})} \varphi\left(\frac{z}{j}\right)\, dz \right\| = \left\| i \int_0^{\eta} e^{2\pi i (R + iy)(\mathbf{A} - s\mathbf{I})} \varphi\left(\frac{R + iy}{j}\right)\, dy \right\|
$$

by (8.9)

$$\leq \int_0^\eta \left\| e^{2\pi i R(\mathbf{A}-s\mathbf{I})} \right\| \left\| e^{-2\pi y(\mathbf{A}-s\mathbf{I})} \right\| \left| \varphi\left(\frac{R+iy}{j} \right) \right| dy$$

by (8.8) with $N = 1$

$$\leq \int_0^\eta e^{2\pi y\|\mathbf{A}-s\mathbf{I}\|} (cj(1+|R+iy|)^{-1} e^{2\pi \frac{y}{j}} \, dy$$

$$\leq cj \int_0^\eta e^{2\pi y(\|\mathbf{A}-s\mathbf{I}\|+\frac{1}{j})} (1+R)^{-1} \, dy$$

$$\leq \frac{Mj e^{2\pi \eta(\|\mathbf{A}-s\mathbf{I}\|+\frac{1}{j})}}{1+R}$$

where M is a constant and the last term clearly tends to 0 if $R \to \infty$.

For the case when $\eta < 0$ the proof is the same and we can conclude that on the vertical segments the integrals are zero. So the integral on the real line and the integral on the complex line $t + i\eta$ are the same except for one sign, so

$$-u_j(s) = \int_{-\infty}^{\infty} e^{2\pi i(t+i\eta)(\mathbf{A}-s\mathbf{I})} \varphi\left(\frac{t+i\eta}{j} \right) \, dt.$$

Now

$$\|u_j(s)\| \leq \int_{-\infty}^{\infty} \left\| e^{2\pi i(t+i\eta)(\mathbf{A}-s\mathbf{I})} \right\| \left| \varphi\left(\frac{t+i\eta}{j} \right) \right| dt$$

from (8.8)

$$\leq c_k j^k \left\| e^{-2\pi \eta(\mathbf{A}-s\mathbf{I})} \right\| e^{\frac{2\pi\eta}{j}} \int_{-\infty}^{\infty} \frac{dt}{(1+|t+i\eta|)^k}$$

$$\leq c_k j^k e^{2\pi(|\eta|\|\mathbf{A}\|-\eta s + \frac{|\eta|}{j})} \int_{-\infty}^{\infty} \frac{dt}{(1+|t|)^k}.$$

In the last inequality we used

$$\|e^{\mathbf{A}-s\mathbf{I}}\| = \|e^{\mathbf{A}} e^{-s\mathbf{I}}\| \leq \|e^{\mathbf{A}}\| \|(e^{\mathbf{I}})^{-s}\| \leq e^{\|\mathbf{A}\|} e^{-s} = e^{\|\mathbf{A}\|-s}.$$

The last integral will converge for $k \geq 2$. Now for each s, we choose $\eta = \alpha \frac{s}{|s|}$ with $\alpha > 0$, the exponential in the last inequality can be written as $e^{2\pi\alpha(\|\mathbf{A}\|-|s|+\frac{1}{j})}$, then $u_j(s) = 0$ if $|s| > \|\mathbf{A}\| + \frac{1}{j}$ and $\alpha \to \infty$. So if we set $M = \|\mathbf{A}\|$ we are done. For that we need to note that if $\psi \in C_0^\infty(\mathbb{R})$ is such that $supp(\psi) \subset \{\xi \in \mathbb{R} : |\xi| > \|\mathbf{A}\|\}$, then $supp(\psi) \subset \{\xi \in \mathbb{R} : |\xi| > \|\mathbf{A}\| + \frac{1}{j}\}$ for large j. $\square$

We have proved that if $\psi \in C_0^\infty$, then $W(\psi, \mathbf{A})$ is a distribution with compact support contained in $\{t : |t| \leq \|\mathbf{A}\|\}$. Another proof can be done using the Paley–Wiener Theorem E.52.

Anderson in [An] proved that the support of this distribution is exactly the spectrum of $\mathbf{A}$.

Let us note that if we take $\phi \in C_0^\infty$ such that $\phi = 1$ on the ball $\{\, t : |t| \leq \|\mathbf{A}\| \,\}$ and we consider now $W(\phi\psi, \mathbf{A})$ where ψ is in C^∞, then $W(\phi\psi, \mathbf{A})$ does not depend on the choice of ϕ since for $\phi_1, \phi_2 \in C_0^\infty$ we have by the previous Lemma, $W((\phi_1 - \phi_2)\psi, \mathbf{A}) = 0$, since $supp\,(\phi_1 - \phi_2) \subset \{\xi \in \mathbb{R} : |\xi| > \|\mathbf{A}\| \,\}$. We can now extend $W(\psi, \mathbf{A})$ to functions ψ in C^∞ by using

$$W(\psi, \mathbf{A}) = W(\phi\psi, \mathbf{A}),$$

where $\phi \in C_0^\infty$ such that $\phi = 1$ on the ball $\{\, t : |t| \leq \|\mathbf{A}\| \,\}$. Now since the polynomials are in C^∞ we are finally in the position to show the correspondence between the Weyl calculus for polynomials and the definition of $p(\mathbf{A})$ given in previous chapters.

Proposition 8.9. *If p is a polynomial the $W(p, \mathbf{A}) = p(\mathbf{A})$.*

Proof.   Given $\varphi \in C_0^\infty$ such that $\varphi = 1$ in a neighborhood of zero, by Lemma 8.4 we have

$$\lim_{j \to \infty} W\left(\varphi\left(\frac{\xi}{j}\right) p(\xi), \mathbf{A}\right) = p(\mathbf{A}).$$

For $\|\mathbf{A}\|$ there is a $j \geq j_0$, (j_0 depending on $\|\mathbf{A}\|$) such that $\varphi\left(\frac{\xi}{j}\right) = 1$ for ξ in a neighborhood of $\{\xi : \xi \leq \|\mathbf{A}\| \,\}$ and by Lemma 8.5

$$W\left(\varphi\left(\frac{\xi}{j}\right) p, \mathbf{A}\right) = W\left(\varphi\left(\frac{\xi}{j_0}\right) p, \mathbf{A}\right) = W(p, A)$$

for every $j \geq j_0$,

and then

$$p(\mathbf{A}) = \lim_{j \to \infty} W\left(\varphi\left(\frac{\xi}{j}\right) p, \mathbf{A}\right) = W(p, \mathbf{A}). \qquad \square$$

Now we are ready to prove that $W(\varphi, \mathbf{A})$ is a functional calculus on $\mathcal{A}$ where $\mathcal{A} \subset L(X)$, X is a complex Hilbert space and $\mathcal{A}$ is the closed self-adjoint operator subalgebra and $C^\infty(\mathbb{R}) = \{\varphi : \mathbb{R} \longrightarrow \mathbb{R} : \varphi$ is infinite differentiable$\}$ with the topology of uniform convergence on compact subsets of functions and their derivatives.

Theorem 8.10. $W : C^\infty(\mathbb{R}) \times \mathcal{A} \to \mathcal{A}$ *is a complete functional calculus, continuous in both variables simultaneously and is an extension of the Riesz functional calculus.*

Proof.   Since the Weyl's functional calculus for polynomials coincide with the Riesz's functional calculus for polynomials as shown in Proposition 8.9 and we have seen that Weyl's functional calculus can be extended to $C^\infty(\mathbb{R})$, we can say that it is an extension of the Riesz functional calculus.

Now given α and β in $C^\infty(\mathbb{R})$ and $\psi \in C_0^\infty$ such that ψ is 1 in a neighborhood of $\{\,\xi : |\xi| \le \|\mathbf{A}\|\,\}$ we have that

$$W(\alpha\beta, \mathbf{A}) = \int_{-\infty}^{\infty} e^{2\pi i t\mathbf{A}}\,\widehat{\psi\alpha\psi\beta}(t)\,dt = \int_{-\infty}^{\infty} e^{2\pi i t\mathbf{A}}\left(\widehat{\psi\alpha} * \widehat{\psi\beta}\right)dt$$

$$= \int_{-\infty}^{\infty} e^{2\pi i t\mathbf{A}}\left(\int_{-\infty}^{\infty} \widehat{\psi\alpha}(s)\cdot\widehat{\psi\beta}(t-s)\,ds\right)dt$$

$$= \int_{-\infty}^{\infty}\int_{-\infty}^{\infty} e^{2\pi i t\mathbf{A}}\widehat{\psi\alpha}(s)\cdot\widehat{\psi\beta}(t-s)\,ds\,dt\,.$$

Using the change of variables $t \longrightarrow r = t-s$ and the fact that the operators $2\pi i s\mathbf{A}$ and $2\pi i r\mathbf{A}$ commute,

$$W(\alpha\beta, \mathbf{A}) = \int_{-\infty}^{\infty}\int_{-\infty}^{\infty} e^{2\pi i(r+s)\mathbf{A}}\widehat{\psi\alpha}(s)\cdot\widehat{\psi\beta}(r)\,ds\,dr$$

$$= \int_{-\infty}^{\infty} e^{2\pi i r\mathbf{A}}\widehat{\psi\beta}(r)\,dr \cdot \int_{-\infty}^{\infty} e^{2\pi i s\mathbf{A}}\widehat{\psi\alpha}(s)\,ds$$

$$= W(\alpha, \mathbf{A})\cdot W(\beta, \mathbf{A}).$$

Since the proof that $W(p\alpha+q\beta, \mathbf{A}) = pW(\alpha, \mathbf{A})+qW(\beta, \mathbf{A})$ with $p, q \in \mathbb{C}$ follows from the properties of the integral, $W(\,\cdot\,, \mathbf{A}) : C^\infty(\mathbb{R})\times \mathcal{A} \longrightarrow \mathcal{A}$ is an homomorphism.

By Proposition 8.9 it is clear that $W(1, \mathbf{A}) = I$ for every $\mathbf{A} \in \mathcal{A}$ and $W(x, \mathbf{A}) = \mathbf{A}$ for every $\mathbf{A} \in \mathcal{A}$. So we have a functional calculus.

To prove that $W(\,\cdot\,,\,\cdot\,)$ is continuous in both variables simultaneously we will calculate for α, $\beta \in C^\infty$, $\mathbf{A}$, $\mathbf{B} \in \mathcal{A}$,

$$\|W(\alpha, \mathbf{A}) - W(\beta, \mathbf{B})\| \le \|W((\alpha - \beta), \mathbf{A})\| + \|W(\beta, \mathbf{A}) - W(\beta, \mathbf{B})\|$$

and will find bounds for each term on the right hand side of the inequality.

Take $\psi \in C_0^\infty$ such that ψ is 1 in a neighborhood of the ball centered in zero with radius max $(\|\mathbf{A}\|, \|\mathbf{B}\|)$. Using Theorem E.20 we get:

$$
\begin{aligned}
\|W(\alpha - \beta, \mathbf{A})\| &= \left\| \int_{-\infty}^{\infty} e^{2\pi i t \mathbf{A}} \overline{\psi(\alpha - \beta)}(t)\, dt \right\| \\
&\leq \int_{-\infty}^{\infty} \left\| e^{2\pi i t \mathbf{A}} \right\| \left| \widehat{\psi(\alpha - \beta)}(t) \right| dt \\
&\leq \int_{-\infty}^{\infty} \left| \widehat{\psi(\alpha - \beta)}(t) \right| dt \\
&\leq C_1 \left\| \left(1 + x^2\right) \widehat{\psi(\alpha - \beta)} \right\|_{\infty} \\
&\leq C_2 \sup_{\substack{x \in supp\ \psi \\ 0 \leq k \leq 2}} \left| \frac{d^k}{dx^k}(\alpha - \beta) \right|
\end{aligned}
\tag{8.10}
$$

which implies continuity of $W(., \mathbf{A})$ using the topology of $\mathcal{C}^{\infty}(\mathbb{R})$, with C_1 and C_2 constants.

To find a bound for

$$
\|W(\beta, \mathbf{A}) - W(\beta, \mathbf{B})\| \leq \int_{-\infty}^{\infty} \left\| e^{2\pi i t \mathbf{A}} - e^{2\pi i t \mathbf{B}} \right\| |\widehat{\psi(\beta)}(t)|\, dt \ ,
$$

where $\psi \in \mathcal{C}_0^{\infty}$ is such that ψ is 1 in a neighborhood of the ball with center 0 and radius max$(\|\mathbf{A}\|, \|\mathbf{B}\|)$, we will prove an interesting lemma which was first proved by L. Tartar and will be very useful in this proof.

Lemma 8.11. $\left\| e^{i\mathbf{A}} - e^{i\mathbf{B}} \right\| \leq \|\mathbf{A} - \mathbf{B}\|$ *if* $\mathbf{A}$ *and* $\mathbf{B}$ *are self-adjoint.*

Proof. We will use the identity

$$
\mathbf{S}^n - \mathbf{T}^n = \sum_{k=0}^{n-1} \mathbf{S}^{n-k-1}(\mathbf{S} - \mathbf{T})\mathbf{T}^k.
$$

Since $e^{i\mathbf{A}}$, $e^{i\mathbf{B}}$ are unitary,

$$\left\| e^{i\mathbf{A}} - e^{i\mathbf{B}} \right\| = \left\| \left(e^{\frac{i\mathbf{A}}{n}} \right)^n - \left(e^{\frac{i\mathbf{B}}{n}} \right)^n \right\|$$

$$= \left\| \sum_{j=0}^{n-1} \left(e^{\frac{i\mathbf{A}}{n}} \right)^{n-j-1} \left(e^{\frac{i\mathbf{A}}{n}} - e^{\frac{i\mathbf{B}}{n}} \right) \left(e^{\frac{i\mathbf{B}}{n}} \right)^j \right\|$$

$$\leq n \left\| e^{\frac{i\mathbf{A}}{n}} - e^{\frac{i\mathbf{B}}{n}} \right\|$$

$$= n \left\| \sum_{k \geq 0} \frac{1}{k!} \left[\left(\frac{i\mathbf{A}}{n} \right)^k - \left(\frac{i\mathbf{B}}{n} \right)^k \right] \right\|$$

$$= n \left\| \sum_{k \geq 1} \frac{1}{k!} \left(\frac{i}{n} \right)^k \sum_{\ell=0}^{k-1} \mathbf{A}^{k-\ell-1}(\mathbf{A} - \mathbf{B})\mathbf{B}^\ell \right\|$$

$$\leq \sum_{k \geq 1} \frac{1}{k!} \left(\frac{1}{n} \right)^{k-1} k C^{k-1} \|\mathbf{A} - \mathbf{B}\|$$

$$= \sum_{k \geq 1} \frac{1}{(k-1)!} \left(\frac{C}{n} \right)^{k-1} \|\mathbf{A} - \mathbf{B}\| = e^{\frac{C}{n}} \|\mathbf{A} - \mathbf{B}\|,$$

where $C = \max(\|\mathbf{A}\|, \|\mathbf{B}\|)$, and if $n \to \infty$ we get the desired inequality. $\square$

Now from Lemma 8.11

$$\|W(\beta, \mathbf{A}) - W(\beta, \mathbf{B})\| \leq \int_{-\infty}^{\infty} \left\| e^{2\pi it\mathbf{A}} - e^{2\pi it\mathbf{B}} \right\| \left| \widehat{\psi(\beta)}(t) \right| dt$$

$$\leq 2\pi \|\mathbf{A} - \mathbf{B}\| \int_{-\infty}^{\infty} |t| \left| \widehat{\psi(\beta)}(t) \right| dt$$

$$\leq 2\pi C \|\mathbf{A} - \mathbf{B}\| \sup_{\substack{x \in supp\, \psi \\ 0 \leq k \leq 3}} \left| \frac{d^k}{dx^k} \beta(x) \right|. \qquad (8.11)$$

Using the estimations (8.10) and (8.11) if $\varphi_n \longrightarrow \varphi$ in $C^\infty(\mathbb{R})$ (uniform convergence of the derivatives in compact sets of $\mathbb{R}$) and if $\mathbf{A}_n \longrightarrow \mathbf{A}$ in $L(X)$ ($\mathbf{A}_n$ and $\mathbf{A}$ self-adjoint), we have

$$W(\varphi_n, \mathbf{A}_n) \longrightarrow W(\varphi, \mathbf{A}) \qquad \text{in } L(X).$$

To prove the completeness we will follow 6.4 in the following sense. Given $\varphi \in C^\infty(\mathbb{R})$ there exists a sequence of polynomials $\{P_k\}$ such that $P_k \longrightarrow \varphi$ in $C^\infty(\mathbb{R})$. Furthermore $P_k \longrightarrow \varphi$ in $C(\mathbb{R})$. So

$$W(\varphi, \mathbf{A}) = \lim_{k \to \infty} W(P_k, \mathbf{A}) = \lim_k P_k(\mathbf{A}) = \varphi(\mathbf{A}).$$

The completeness clearly follows from what was done for continuous functions in Theorem 6.4. $\square$

For the definition of $W(\varphi, \mathbf{A})$ and the continuity in both variables it is not necessary for φ to be real valued, it can be a complex valued function and we will have:

Proposition 8.12.

$$W(\varphi, \mathbf{A})^* = W(\overline{\varphi}, \mathbf{A}) ,$$

where $\varphi : \mathbb{R} \longrightarrow \mathbb{C}$, $\varphi \in C^\infty$, *and* $W(\varphi, \mathbf{A})^*$ *denotes the adjoint of* $W(\varphi, \mathbf{A})$.

Proof. The map $\mathbf{A} \longrightarrow \mathbf{A}^*$ from $L(X) \longrightarrow L(X)$ is a continuous operator (Remark C.1) Using Proposition 1.4 we can write the following:
Let us take $\varphi \colon \mathbb{R} \to \mathbb{C}$, $\varphi \in C^\infty$,

$$\begin{aligned}
W(\varphi, \mathbf{A})^* &= \left[\int_{-\infty}^{\infty} e^{2\pi i t \mathbf{A}} \left(\widehat{\psi\varphi} \right)(t)\, dt \right]^* \\
&= \int_{-\infty}^{\infty} \left(e^{2\pi i t \mathbf{A}} \right)^* \overline{\left(\widehat{\psi\varphi} \right)(t)}\, dt \\
&= \int_{-\infty}^{\infty} e^{-2\pi i t \mathbf{A}} \overline{\left(\widehat{\psi\varphi} \right)(t)}\, dt \\
&= -\int_{\infty}^{-\infty} e^{2\pi i t \mathbf{A}} \overline{\left(\widehat{\psi\varphi} \right)(-t)}\, dt \\
&= \int_{-\infty}^{\infty} e^{2\pi i t \mathbf{A}} \left(\widehat{\psi\overline{\varphi}} \right)(t)\, dt = W(\overline{\varphi}, \mathbf{A}) .
\end{aligned} \tag{8.12}$$

We can take ψ to be real and then we can apply the following to get the equality (8.12).

$$\overline{\left(\widehat{\psi\varphi} \right)}(-t) = \overline{\mathscr{F} \left(\widehat{\psi\varphi} \right)}(-t) = \overline{\mathscr{F}}(\psi\varphi)(-t) = \widehat{\psi\overline{\varphi}}(t) . \qquad \square$$

Now we will extend the formula (8.3) to the multi-variable case, even if the operators involved are not commutative.
Let X be a Hilbert space, and fix $\{ \mathbf{A}_1, \dots, \mathbf{A}_n \} \subset L(X)$ with each $\mathbf{A}_i$ self-adjoint. For $\xi = (\xi_1, \dots, \xi_n) \in \mathbb{R}^n$, we consider the map $\xi \longrightarrow e^{2\pi i \xi \mathbf{A}}$, where we use $\xi \mathbf{A}$ to denote the sum $\xi_1 \mathbf{A}_1 + \xi_2 \mathbf{A}_2 + \cdots + \xi_n \mathbf{A}_n$. Observe that the operator $e^{2\pi i \xi \mathbf{A}}$ is unitary for each ξ. Now let $f \in \mathcal{S}(\mathbb{R}^n)$ and consider the formal Bochner integral:

$$W(f, \mathbf{A}) = \int_{\mathbb{R}^n} e^{2\pi i \xi \mathbf{A}} \widehat{f}(\xi)\, d\xi .$$

If the $\mathbf{A}_i$ commute then we could apply the ideas from the beginning of the chapter to obtain $W(f, \mathbf{A}) = f(\mathbf{A})$ in the sense of spectral calculus.

However, since this is not assumed we must consider the nature of the operator valued distribution $W(f, \mathbf{A})$ more carefully.

In this case we will follow the lines of what was already done for one operator. We will start with an inequality ([Ne]) needed to extend Lemma 8.5.

Lemma 8.13. *Let* $\mathbf{A}, \mathbf{B} \in L(X)$ *two self-adjoint operators then:*

$$\left\| e^{\mathbf{A}+i\mathbf{B}} \right\| \leq \left\| e^{\mathbf{A}} \right\|.$$

Proof. For $b > \|\mathbf{A}\|$ consider the operator $\mathbf{A}_b = \mathbf{A} - b\mathbf{I}$. Then

$$\left\| e^{\mathbf{A}_b} \right\| = e^{-b} \left\| e^{\mathbf{A}} \right\| \leq e^{-b} e^{\|\mathbf{A}\|} = e^{-b+\|\mathbf{A}\|} \leq 1.$$

We claim that

$$\lim_{n \to \infty} \left(e^{\frac{\mathbf{A}_b}{n}} e^{\frac{i\mathbf{B}}{n}} \right)^n = e^{\mathbf{A}_b + i\mathbf{B}}.$$

We need to calculate

$$\left(e^{\frac{\mathbf{A}_b}{n}} e^{\frac{i\mathbf{B}}{n}} \right)^n - e^{\mathbf{A}_b + i\mathbf{B}} = \left(e^{\frac{\mathbf{A}_b}{n}} e^{\frac{i\mathbf{B}}{n}} \right)^n - \left(e^{\frac{\mathbf{A}_b + i\mathbf{B}}{n}} \right)^n.$$

We can write, using $\mathbf{M} = e^{\frac{\mathbf{A}_b}{n}} e^{\frac{i\mathbf{B}}{n}}$ and $\mathbf{N} = e^{\frac{\mathbf{A}_b + i\mathbf{B}}{n}}$,

$$\mathbf{M}^n - \mathbf{N}^n = \sum_{k=0}^{n-1} \mathbf{M}^{n-k-1}(\mathbf{M} - \mathbf{N})\mathbf{N}^k$$

and since $\|\mathbf{M}\| \leq 1$, we can estimate:

$$\|\mathbf{M}^n - \mathbf{N}^n\| \leq \sum_{k=0}^{n-1} \|\mathbf{M} - \mathbf{N}\| \, \|\mathbf{N}^k\| \leq n e^{\|\mathbf{A}_b + i\mathbf{B}\|} \|\mathbf{M} - \mathbf{N}\|. \qquad (8.13)$$

We also have:

$$\mathbf{M} - \mathbf{N} = \sum_{\ell \geq 0} \left(\frac{\mathbf{A}_b}{n} \right)^\ell \frac{1}{\ell!} \sum_{h \geq 0} \left(\frac{i\mathbf{B}}{n} \right)^h \frac{1}{h!} - \sum_{k \geq 0} \left(\frac{\mathbf{A}_b + i\mathbf{B}}{n} \right)^k \frac{1}{k!}$$

$$= \sum_{k \geq 0} \frac{1}{n^k} \left[\sum_{\ell+h=k} \frac{1}{\ell! h!} \mathbf{A}_b^\ell (i\mathbf{B})^h - \frac{(\mathbf{A}_b + i\mathbf{B})^k}{k!} \right]$$

$$= \frac{1}{n^2} \sum_{k \geq 0} \frac{1}{n^k} \left[\sum_{h+\ell=k+2} \frac{1}{h! \ell!} \mathbf{A}_b^\ell (i\mathbf{B})^h - \frac{(\mathbf{A}_b + i\mathbf{B})^{k+2}}{(k+2)!} \right].$$

Since the terms corresponding to $k = 0$, $k = 1$ are zero, taking norms:

$$\|\mathbf{M} - \mathbf{N}\| \leq \frac{1}{n^2} \sum_{k \geq 0} \frac{1}{n^k} \left[\sum_{h+\ell=k+2} \frac{1}{h!} \frac{1}{\ell!} \|\mathbf{A}_b\|^\ell \|\mathbf{B}\|^h + \frac{\|\mathbf{A}_b + i\mathbf{B}\|^{k+2}}{(k+2)!} \right]$$

$$= \frac{1}{n^2} C\left(\|\mathbf{A}_b\|, \|\mathbf{B}\| \right).$$

(constant C times a function of $\|\mathbf{A}_b\|$ and $\|\mathbf{B}\|$).

Now using inequality (8.13):

$$\left\| (e^{\frac{\mathbf{A}_b}{n}} e^{\frac{i\mathbf{B}}{n}})^n - e^{\mathbf{A}_b + i\mathbf{B}} \right\| \leq \frac{1}{n} e^{\|\mathbf{A}_b + i\mathbf{B}\|} C(\|\mathbf{A}_b\|, \|\mathbf{B}\|)$$

and the last part converges to zero if n converges to infinity. Then the claim is proved and

$$\lim_{n \to \infty} \left\| \left(e^{\frac{\mathbf{A}_b}{n}} e^{\frac{i\mathbf{B}}{n}} \right)^n \right\| = \left\| e^{\mathbf{A}_b + i\mathbf{B}} \right\| \leq \overline{\lim_{n \to \infty}} \left\| e^{\frac{\mathbf{A}_b}{n}} \right\|^n . \tag{8.14}$$

Since $e^{\frac{\mathbf{A}_b}{n}}$ is self-adjoint we have that

$$\left\| e^{\mathbf{A}_b} \right\| = \left\| e^{\frac{\mathbf{A}_b}{2}} e^{\frac{\mathbf{A}_b}{2}} \right\| = \left\| e^{\frac{\mathbf{A}_b}{2}} \right\|^2 = \left\| e^{\frac{\mathbf{A}_b}{4}} \right\|^4 = \cdots = \left\| e^{\frac{\mathbf{A}_b}{2^k}} \right\|^{2^k} . \tag{8.15}$$

By (8.14) and (8.15) it follows:

$$\left\| e^{\mathbf{A}_b + i\mathbf{B}} \right\| \leq \lim_k \left\| e^{\frac{\mathbf{A}_b}{2^k}} \right\|^{2^k} = \left\| e^{\mathbf{A}_b} \right\|$$

or

$$\left\| e^{\mathbf{A} - b\mathbf{I} + i\mathbf{B}} \right\| \leq \left\| e^{\mathbf{A} - b\mathbf{I}} \right\|$$

and

$$e^{-b} \left\| e^{\mathbf{A} + i\mathbf{B}} \right\| \leq e^{-b} \left\| e^{\mathbf{A}} \right\| .$$

Then $\left\| e^{\mathbf{A} + i\mathbf{B}} \right\| \leq \left\| e^{\mathbf{A}} \right\|$. $\qquad\square$

Remark 8.14. For every $N \in \mathbb{N}$, $\sum_{k=0}^{N} \frac{(2\pi i)^k}{k!} (\xi_1 \mathbf{A}_1 + \xi_2 \mathbf{A}_2 + \cdots + \xi_n \mathbf{A}_n)^k$ is a polynomial in $(\xi_1, \ldots, \xi_n)$, so it is an entire function and as $N \to \infty$, the polynomials converge uniformly on compact sets to $e^{2\pi i \xi \mathbf{A}}$, and so it is also an entire function.

Proposition 8.15. *Let* $\mathbf{A} = (\mathbf{A}_1, \ldots, \mathbf{A}_n)$ *where* $\mathbf{A}_k \in L(X)$ *and* $\mathbf{A}_k$ *is self-adjoint. If* $\psi \in C_0^\infty$ *is such that* supp $\psi \subset \bigcup_k \{ |t_k| > \|\mathbf{A}_k\| \}$, *then the Bochner integral:*

$$W(\psi, \mathbf{A}) = \int_{\mathbb{R}^n} e^{2\pi i \xi \mathbf{A}} \widehat{\psi}(\xi) \, d\xi = 0 .$$

Proof.   Suppose the operators $\mathbf{A}_k$ commute.
Using the notation of Lemma 8.5 we can write

$$u_j(s) = \int_{\mathbb{R}^n} e^{2\pi i \xi (\mathbf{A} - s\mathbf{I})} \varphi\left(\frac{\xi_1}{j}\right) \cdots \varphi\left(\frac{\xi_n}{j}\right) d\xi$$

$$= \prod_{\ell=1}^{n} \int_{-\infty}^{\infty} e^{2\pi i \xi_\ell (\mathbf{A}_\ell - s_\ell \mathbf{I})} \varphi\left(\frac{\xi_\ell}{j}\right) d\xi_\ell = \prod_{\ell=1}^{n} u_j^{(\ell)}(s).$$

If the operators $\mathbf{A}_k$ commute, $u_j(s)$ is a function of separate variables
and we may apply in each component Lemma 8.5 and as a result we get
$u_j(s) = 0$ if $|s_\ell| > \mathbf{A}_\ell$ for all $\ell = 1, \ldots, n$. Then we follow the proof of
Lemma 8.5 and we have that $W(\psi, \mathbf{A}) = 0$.

If operators $\mathbf{A}_1, \ldots, \mathbf{A}_n$ do not commute we can not be sure that $u_j(s)$
is a function of separate variables. Let's consider again

$$u_j(s) = \int_{\mathbb{R}^n} e^{2\pi i \xi (\mathbf{A} - s\mathbf{I})} \varphi\left(\frac{\xi_1}{j}\right) \cdots \varphi\left(\frac{\xi_n}{j}\right) d\xi.$$

Take a vector of the form $e_k = (0, \ldots, 0, 1, 0, \ldots, 0)$ and consider ηe_k for
$\eta \in \mathbb{R}$.

Claim 8.16.

$$u_j(s) = \int_{\mathbb{R}^n} e^{2\pi i \xi (\mathbf{A} - s\mathbf{I})} \varphi\left(\frac{\xi_1}{j}\right) \cdots \varphi\left(\frac{\xi_n}{j}\right) d\xi$$

$$= \int_{\mathbb{R}^n} e^{2\pi i (\xi + i\eta e_k)(\mathbf{A} - s\mathbf{I})} \varphi\left(\frac{\xi_1}{j}\right) \cdots \varphi\left(\frac{\xi_k + i\eta}{j}\right) \cdots \varphi\left(\frac{\xi_n}{j}\right) d\xi.$$

$$\text{(8.16)}$$

Since the integral exists, by Fubini, it is enough to prove for
$\xi_1, \ldots, \xi_{k-1}, \xi_{k+1}, \ldots, \xi_n$ fixed the equality:

$$\int_{-\infty}^{\infty} e^{2\pi i (\xi + i\eta e_k)(\mathbf{A} - s\mathbf{I})} \varphi\left(\frac{\xi_k + i\eta}{j}\right) d\xi_k = \int_{-\infty}^{\infty} e^{2\pi i \xi (\mathbf{A} - s\mathbf{I})} \varphi\left(\frac{\xi_k}{j}\right) d\xi_k.$$

In the same way we did in Lemma 8.5 here we will also use Cauchy's
Theorem in the rectangle with vertices R, $R + i\eta$, $R - i\eta$, $-R$ assuming R
and η are positive. Consider the entire function:

$$F(z) = e^{2\pi i \xi'(\mathbf{A}' - s'\mathbf{I}) + 2\pi i z (\mathbf{A}_k - s_k \mathbf{I})} \varphi\left(\frac{z}{j}\right)$$

where the primes denote the $n - 1$-tuples where the k-th coordinate is
missing and $z = \xi_k + iy$.

Now, we want to prove that for $R \longrightarrow \infty$ the integrals on the vertical
lines tend to zero.

We will use the previous remark and the bound calculated in inequality (8.8):

$$\left| \varphi \left(\frac{z}{j} \right) \right| \le c_k j^k (1 + |z|)^{-k} e^{\frac{2\pi |y|}{j}} .$$

Take $\eta > 0$,

$$\left\| \int_0^\eta F(R + iy) i \, dy \right\| \le \int_0^\eta \| F(R + iy) i \| \, dy \le \int_0^\eta e^{2\pi |y| \|\mathbf{A}_k - s_k\|} \left| \varphi \left(\frac{z}{j} \right) \right| dy$$

$$\le \int_0^\eta e^{2\pi \eta \|\mathbf{A}_k - s_k \mathbf{I}\|} c_1 j (1 + |z|)^{-1} e^{\frac{2\pi |y|}{j}} \, dy$$

$$\le \int_0^\eta e^{2\pi \eta \|\mathbf{A}_k - s_k \mathbf{I}\|} c \frac{1}{1 + R} e^{2\pi \eta} \, dy$$

$$\le c \, e^{2\pi \eta \|\mathbf{A}_k - s_k \mathbf{I}\|} e^{2\pi \eta} \frac{\eta}{1 + R}$$

which tends to 0 as R tends to ∞, and the same is true for $\eta < 0$. So (8.16) holds.

The proof continues in the following way:

$$\| u_j(s) \| \le \int_{\mathbb{R}^n} \left\| e^{2\pi i (\xi + i\eta e_k)(\mathbf{A} - s\mathbf{I})} \varphi \left(\frac{\xi_1}{j} \right) \cdots \varphi \left(\frac{\xi_k + i\eta}{j} \right) \cdots \varphi \left(\frac{\xi_n}{j} \right) \right\| d\xi$$

$$\le c_{\ell,k} \, j^{\ell k} \left\| e^{-2\pi \eta (\mathbf{A}_k - s_k \mathbf{I})} \right\| e^{\frac{2\pi |\eta|}{j}} \int_{\mathbb{R}^n} \frac{d\xi}{(1 + |\xi + i\eta|)^\ell}$$

$$\le c_{\ell,k} \, j^{\ell k} e^{2\pi \left(|\eta| \|\mathbf{A}_k - \eta s_k + \frac{|\eta|}{j} \right)} \int_{\mathbb{R}^n} \frac{d\xi}{(1 + |\xi|)^\ell} .$$

The integral will converge if $\ell \ge 2$.

For each s we choose $\eta = \alpha \dfrac{s_k}{|s_k|}$ with $\alpha > 0$. The exponential in the last inequality can be written as:

$$e^{2\pi \alpha \left(\|\mathbf{A}_k\| - |s_k| + \frac{1}{j} \right)} .$$

Then

$$u_j(s) = 0 \quad \text{if} \quad |s_k| > \|\mathbf{A}_k\| + \frac{1}{j} \quad \text{and} \quad \alpha \longrightarrow \infty ,$$

and

$$W(\psi, \mathbf{A}) = \lim_{j \to \infty} \int_{\mathbb{R}^n} u_j(s) \, \psi(s) \, ds = 0 \quad \text{if}$$

$$\psi \in C_0^\infty \quad \text{and} \quad supp \, \psi \subset \bigcup_{k=1}^n \{ |s_k| > \|\mathbf{A}_k\| \} .$$

$\square$

We have proved that if $\psi \in C_0^\infty$, then $W(\psi, \mathbf{A})$ is a distribution with compact support contained in $\bigcap_{k=1}^n \{ |s_k| > \|\mathbf{A}_k\| \}$. By using [Ho](4-2-3) or [Ru2](6.24) we can extend $W(\psi, \mathbf{A})$ to functions $\psi \in C^\infty$ (see Appendix E).

Given a function $\varphi \in C^\infty(\mathbb{R}^n)$ it can happen that the function depends on less than the n variables $\xi_1, \ldots, \xi_n$; if ξ' indicates the group of variables on which the function depends, we can consider the expression $W(\varphi(\xi), \mathbf{A})$ where we use a truncation and an integral in all variables ξ. We can also consider the expression and $W(\varphi(\xi'), \mathbf{A}')$ Where $\mathbf{A}'$ is the set of operators corresponding to the variable ξ'.
We will prove that both expressions coincide so there is compatibility of the different dimensions.

Proposition 8.17. *Let $\xi \in \mathbb{R}^n$, $\xi = (\xi', \xi'')$ where ξ' is a k-tuple, $k < n$, formed with k of the variables $\xi_1, \ldots, \xi_n$. Let $(\mathbf{A}_1, \ldots, \mathbf{A}_n) \in \mathcal{A}^n$ and let's denote by $(\mathbf{A}', \mathbf{A}'')$ the corresponding operators. If $\varphi(\xi) \in C^\infty(\mathbb{R}^n)$ and $\varphi(\xi) = \varphi(\xi') \ \big(\varphi(\xi') \in C^\infty(\mathbb{R}^k) \big)$, then*

$$W\left(\varphi(\xi), \mathbf{A}\right) = W\left(\varphi(\xi'), \mathbf{A}'\right).$$

Proof. Take $\psi \in C_0^\infty(\mathbb{R})$ such that $\psi = 1$ on a neighborhood of zero. Then set $\psi(t) = \psi(t_1) \ldots \psi(t_n)$ and let $\psi(t')$ and $\psi(t'')$ the corresponding functions and $j \in \mathbb{N}$. Then

$$W(\varphi, \mathbf{A}) = \int_{\mathbb{R}^n} e^{2\pi i \xi \mathbf{A}} \left[\widehat{\varphi(s')\psi(s')\psi\left(\frac{s''}{j}\right)} \right](\xi) \, d\xi$$

$$= \int_{\mathbb{R}^k} \int_{\mathbb{R}^{n-k}} e^{2\pi i (\xi' \mathbf{A}' + \xi'' \mathbf{A}'')} \left[\widehat{\varphi(s')\psi(s')\psi\left(\frac{s''}{j}\right)} \right](\xi', \xi'') \, d\xi' \, d\xi''$$

$$= \int_{\mathbb{R}^k} \left(\widehat{\varphi(s')\psi(s')} \right)(\xi') \left(\int_{\mathbb{R}^{n-k}} e^{2\pi(\xi' \mathbf{A}' + \xi'' \mathbf{A}'')} \left(\widehat{\psi\left(\frac{s''}{j}\right)} \right)(\xi'') \, d\xi'' \right) d\xi'$$

$$= \int_{\mathbb{R}^k} \left(\widehat{\varphi(s')\psi(s')} \right)(\xi') \left(\int_{\mathbb{R}^{n-k}} e^{2\pi i (\xi' \mathbf{A}' + \xi'' \mathbf{A}'')} j^{n-k} \widehat{\psi(js'')}(\xi'') \, d\xi'' \right) d\xi'.$$

We will study the second integral, the integral on $\mathbb{R}^{n-k}$. First we do a change of variable $j\xi'' = r''$:

$$\int_{\mathbb{R}^{n-k}} e^{2\pi i \left(\xi' \mathbf{A}' + \frac{r''}{j} \mathbf{A}''\right)} \widehat{\psi}(r'') \, dr''.$$

Since $e^{2\pi i \xi \mathbf{A}}$ is unitary, the integrand is bounded by $|\widehat{\psi}(r'')|$ which is independent of j. Furthermore, $e^{2\pi i \left(\xi' \mathbf{A}' + \frac{r''}{j} \mathbf{A}'' \right)} \longrightarrow e^{2\pi i \xi' \mathbf{A}'}$ in $L(X)$ as $j \longrightarrow \infty$. So, using the Dominated Convergence Theorem for Bochner integrals we have:

$$\int_{\mathbb{R}^{n-k}} e^{2\pi i \left(\xi' \mathbf{A}' + \frac{r''}{j} \mathbf{A}'' \right)} \widehat{\psi}(r'') \, dr'' \longrightarrow e^{2\pi i \xi' \mathbf{A}'} \quad \text{in } L(X)$$

since $\psi(0) = 1$.

On the other hand

$$\left\| \int_{\mathbb{R}^{n-k}} e^{2\pi i \left(\xi' \mathbf{A}' + \frac{r''}{j} \mathbf{A}'' \right)} \widehat{\psi}(r'') \, dr'' \right\| \leq \int_{\mathbb{R}^{n-k}} \left| \widehat{\psi}(r'') \right| \, dr'' \, .$$

So we can apply again the dominated convergence theorem, this time in $\mathbb{R}^k$ and we will have that

$$\int_{\mathbb{R}^k} \widehat{\varphi \psi}(\xi') \int_{\mathbb{R}^{n-k}} e^{2\pi i \left(\xi' \mathbf{A}' + \frac{r''}{j} \mathbf{A}'' \right)} \widehat{\psi}(r'') \, dr'' d\xi' \longrightarrow \int_{\mathbb{R}^k} \widehat{\varphi \psi}(\xi')^{2\pi i \xi' \mathbf{A}'} d\xi' \, .$$

So it is clear that

$$W(\varphi(\xi), \mathbf{A}) = W(\varphi(\xi'), \mathbf{A}') . \qquad \square$$

Corollary 8.18. *If $p(t_k)$ is a polynomial in the variable t_k then*

$$W(p, \mathbf{A}) = W(p, \mathbf{A}_k) = p(\mathbf{A}_k) \, .$$

Proof. It is immediate from Propositions 8.9 and 8.17. $\qquad \square$

Let us look now at the polynomial situation. The Weyl functional calculus is a way of constructing functions $F(\mathbf{A}_1, \ldots, \mathbf{A}_n)$ of the n-tuple $(\mathbf{A}_1, \ldots, \mathbf{A}_n)$ of self adjoint operators, where the operators do not necessarily commute with each other and so there is not a unique way of forming such functions. However, the Weyl functional calculus is defined in such a way that there will be a unique way to get such functions. For polynomials the essential problem is for instance if $p(x_1, x_2) = x_1 x_2$, then $p(\mathbf{A}_1, \mathbf{A}_2)$ can be formed with $\mathbf{A}_1 \mathbf{A}_2$ or $\mathbf{A}_2 \mathbf{A}_1$ or $\frac{1}{2}(\mathbf{A}_1 \mathbf{A}_2 + \mathbf{A}_2 \mathbf{A}_1)$ and we have here several choices. If the operators were commutative either one of the possibilities above would give the same result. To consider the non-commuting case, we first consider decomposing a monomial $p^\alpha(x_1, \ldots, x_n) = x_1^{\alpha_1} x_2^{\alpha_2} \cdots x_n^{\alpha_n}$, where $\alpha = (\alpha_1, \ldots, \alpha_n) \in \mathbb{N}^n$ is a multi-index with $|\alpha| = \sum_{i=1}^n \alpha_i$ in a symmetric way.

Define

$$S_\alpha = \{ \pi \colon \{1, 2, \ldots, |\alpha|\} \} \longrightarrow \{ \{1, \ldots, n\} \colon \pi^{-1}(\{j\}) = \alpha_j, \ j \in \{1, \ldots, n\} \} \, .$$

That is every map $\pi \in S_\alpha$ attains the value j exactly α_j times. Using this terminology we have:

$$p^\alpha(x_1, \ldots, x_n) = \frac{\alpha_1! \cdots \alpha_n!}{|\alpha|} \sum_{\pi \in S_\alpha} x_{\pi(1)} \cdots x_{\pi(|\alpha|)}.$$

With this in mind we can relax the commuting conditions on $(\mathbf{A}_1, \ldots, \mathbf{A}_n)$ and define the symmetric operator as:

$$p^\alpha(\mathbf{A}_1, \ldots, \mathbf{A}_n) = \frac{\alpha_1! \cdots \alpha_n!}{|\alpha|} \sum_{\pi \in S_\alpha} \mathbf{A}_{\pi(1)}, \ldots, \mathbf{A}_{\pi(n)}$$

and then there is no ambiguity in defining an operator for polynomials since once we have it defined for p^α it is easy to extend it to every polynomial since the set of polynomials in $\mathbb{R}^n$, $\mathscr{P}(\mathbb{R}^n) = \text{span}\{p^\alpha : \alpha \in \mathbb{N}^n\}$. As an example take $n = 2$, $\mathbf{A} = (\mathbf{A}_1, \mathbf{A}_2)$, $p^3(x_1, x_2) = x_1 x_2^2$. We have $\alpha = (1, 2)$ so $|\alpha| = 3$, $\pi\colon \{1, 2, 3\} \longrightarrow \{1, 2\}$ and we have three maps in S_3, $\pi_1(1) = 1$, $\pi_1(2) = \pi_1(3) = 2$, $\pi_2(1) = \pi_2(3) = 2$, $\pi_2(2) = 1$ and $\pi_3(1) = \pi_3(2) = 2$, $\pi_3(3) = 1$.
Then

$$\begin{aligned} p^{(1,2)}(\mathbf{A}_1, \mathbf{A}_2) &= \frac{1!2!}{3!} \left(\mathbf{A}_1 \mathbf{A}_2 \mathbf{A}_2 + \mathbf{A}_2 \mathbf{A}_1 \mathbf{A}_2 + \mathbf{A}_2 \mathbf{A}_2 \mathbf{A}_1 \right) \\ &= \frac{1}{3} \left(\mathbf{A}_1 \mathbf{A}_2^2 + \mathbf{A}_2 \mathbf{A}_1 \mathbf{A}_2 + \mathbf{A}_2^2 \mathbf{A}_1 \right). \end{aligned}$$

If $\mathbf{A}_1$ and $\mathbf{A}_2$ commute then $p^{(1,2)}(\mathbf{A}_1, \mathbf{A}_2) = \mathbf{A}_1 \mathbf{A}_2^2$. Now we can extend this definition to the general case where p is a polynomial and the operators $(\mathbf{A}_1, \ldots, \mathbf{A}_n)$ are not commutative and it will be consistent with the commutative case.

Furthermore, for each monomial $p^\alpha(x_1, \ldots, x_n) = x_1^{\alpha_1} \cdots x_n^{\alpha_n}$ we define the differential operator

$$\mathbf{D}^\alpha = \left(\frac{1}{2\pi i} \right)^{|\alpha|} \frac{\partial^{\alpha_1}}{\partial x_1^{\alpha_1}} \frac{\partial^{\alpha_2}}{\partial x_2^{\alpha_2}} \cdots \frac{\partial^{\alpha_n}}{\partial x_n^{\alpha_n}}.$$

Now the definition for $p(\mathbf{D})$, where p is a polynomial, follows naturally.

Proposition 8.19. *If p is a polynomial*

$$W(p, \mathbf{A}) = p(\mathbf{D}) \left(e^{2\pi i t \mathbf{A}} \right) \Big|_{t=0} = p(\mathbf{A}).$$

Proof. We will only prove the proposition for a monomial p^α of the form $p^\alpha(t) = t_1^{\alpha_1} \cdots t_n^{\alpha_n} = t^\alpha$.

Take $\psi \in \mathcal{C}_0^\infty$ such that $\psi(0) = 1$.

$$W(t^\alpha, \mathbf{A}) = \int_{\mathbb{R}^n} e^{2\pi i t \cdot \mathbf{A}} \widehat{t^\alpha \psi}\left(\frac{t}{j}\right) dt$$

$$= \int_{\mathbb{R}^n} e^{2\pi i t \cdot \mathbf{A}} \mathbf{D}^\alpha \widehat{\psi}(jt) j^n \, dt .$$

To evaluate this integral we will first prove that the integrand is bounded and then we will use integration by parts. An application of the Cauchy integral formula for several variables gives:

$$\left(\frac{\partial^{\beta_1}}{\partial t_1^{\beta_1}} \cdots \frac{\partial^{\beta_n}}{\partial t_n^{\beta_n}}\right)\left(e^{2\pi i t \cdot \mathbf{A}}\right)$$

$$= \frac{\beta_1! \cdots \beta_n!}{(2\pi i)^n} \int_{|z_1 - t_1| = 1} \cdots \int_{|z_n - t_n| = 1} \frac{e^{2\pi i z \cdot \mathbf{A}}}{(z_1 - t_1)^{\beta_1 + 1} \cdots (z_n - t_n)^{\beta_n + 1}} \, dz .$$

If $z_k = x_k + i y_k$, since $|z_k - t_k| = 1$ and $t_k \in \mathbb{R}$ we have that $|y_k| < 1$ so

$$\left\| \int_{|z_1 - t_1| = 1} \cdots \int_{|z_n - t_n| = 1} \frac{e^{2\pi i z \cdot \mathbf{A}}}{(z_1 - t_1)^{\beta_1 + 1} \cdots (z_n - t_n)^{\beta_n + 1}} \, dz \right\|$$

$$\leq (2\pi)^n e^{2\pi(|y_1| \|\mathbf{A}_1\| + \cdots + |y_n| \|\mathbf{A}_n\|)}$$

$$\leq (2\pi)^n e^{2\pi(\|\mathbf{A}_1\| + \cdots + \|\mathbf{A}_n\|)} .$$

Then

$$\left\| \left(\frac{\partial^{\beta_1}}{\partial t_1^{\beta_1}} \cdots \frac{\partial^{\beta_n}}{\partial t_n^{\beta_n}}\right)\left(e^{2\pi i t \cdot \mathbf{A}}\right) \right\| \leq \beta_1! \cdots \beta_n! e^{2\pi(\|\mathbf{A}_1\| + \cdots + \|\mathbf{A}_n\|)}$$

which proves that the integrand is bounded. So we can now use integration by parts, and we get

$$W(t^\alpha, \mathbf{A}) = \int_{\mathbb{R}^n} \mathbf{D}^\alpha e^{2\pi i t \cdot \mathbf{A}} \widehat{\psi}(jt) j^n \, dt .$$

Now following the same argumentation as used in Lemma 8.4 we can prove the existence of the following limit:

$$\lim_{j \to \infty} \int_{\mathbb{R}^n} \mathbf{D}^\alpha e^{2\pi i t \cdot \mathbf{A}} \widehat{\psi}(jt) j^n \, dt = \mathbf{D}^\alpha e^{2\pi i t \cdot \mathbf{A}} \Big|_{t=0} .$$

From the analyticity of $e^{2\pi it\cdot\mathbf{A}}$ we may calculate:

$$
\begin{aligned}
\mathbf{D}^\alpha \left(e^{2\pi it\cdot\mathbf{A}} \right)\Big|_{t=0} &= \frac{(2\pi i)^{|\alpha|}}{|\alpha|}\mathbf{D}^\alpha (t\cdot\mathbf{A})^{|\alpha|}\Big|_{t=0} \\
&= \frac{(2\pi i)^{|\alpha|}}{|\alpha|}\mathbf{D}^\alpha t^\alpha \sum_{\pi\in S_\alpha} \mathbf{A}_{\pi(1)}\cdots\mathbf{A}_{\pi(\alpha)} \\
&= \frac{\alpha_1!\cdots\alpha_n!}{|\alpha|!}\sum_{\pi\in S_\alpha} \mathbf{A}_{\pi(1)}\cdots\mathbf{A}_{\pi(\alpha)} \\
&= p(\mathbf{A}) \ .
\end{aligned}
$$

The result for arbitrary polynomials follows immediately. $\qquad\square$

Proposition 8.20. *Let $\mathcal{A}$ be a real algebra, closed and self-adjoint in $L(X)$. Then*

$$
W: \mathcal{C}^\infty \times \mathcal{A}^n \longrightarrow L(X)
$$
$$
(\varphi,\mathbf{A}) \longrightarrow W(\varphi,\mathbf{A})
$$

where $\mathbf{A} = (\mathbf{A}_1,\ldots,\mathbf{A}_n)$ is continuous in both variables simultaneously.

Proof. To check the continuity in this case we will follow what was done in Proposition 6.4. Given $\mathbf{A}$, $\mathbf{B}$ n-tuples of operators and $\alpha,\beta\in\mathcal{C}^\infty$. We can write

$$
\|W(\alpha,\mathbf{A}) - W(\beta,\mathbf{A})\| \le \|W(\alpha-\beta,\mathbf{A})\| + \|W(\beta,\mathbf{A}) - W(\beta,\mathbf{B})\| \ .
$$

If $\psi\in\mathcal{C}_0^\infty$ is 1 in a neighborhood of the ball centered at zero and with radius $\|\mathbf{A}\|$, the function in the first term will be bounded by:

$$
\|W(\alpha-\beta,\mathbf{A})\| \le \int_{\mathbb{R}^n} \left|\widehat{\psi(\alpha-\beta)}\right|(\xi)\,d\xi \le C \sup_{\substack{x\in supp\ \psi \\ |\alpha|\le n+1}} |D^\alpha(\alpha-\beta)(x)| \ .
$$

The second term is now bounded by:

$$
\begin{aligned}
\|W(\beta,\mathbf{A}) - W(\beta,\mathbf{B})\| &\le \int_{\mathbb{R}^n} \|2\pi\xi(\mathbf{A}-\mathbf{B})\|\left\|\widehat{\psi\beta(\xi)}\right\|\,d\xi \\
&\le 2\pi c\|\mathbf{A}-\mathbf{B}\| \sup_{\substack{x\in supp\ \psi \\ |\alpha|\le n+2}} |D^\alpha\beta(x)| \ ,
\end{aligned}
$$

where Lemma 8.11 was used and $\|\mathbf{A}-\mathbf{B}\| = \left(\sum_{j=1}^n \|\mathbf{A}_j-\mathbf{B}_j\|^2\right)^{1/2}$. The above estimation proves the continuity in both variables. $\qquad\square$

Remark 8.21. The Weyl functional calculus for a family of self-adjoint operators acting on a Hilbert space provides a map from the space of C^∞-functions on $\mathbb{R}^n$ into the set of bounded operators. The Weyl functional calculus is not multiplicative under pointwise multiplication of functions unless the self-adjoint operators commute.

There are various approaches to construct a functional calculus of self-adjoint operators. One possibility would be to generalize the Riesz functional calculus to higher dimensions. The problem with this approach is the need of the Cauchy integral. The Clifford algebras have developed a higher dimensional Clifford–Cauchy integral. The Clifford approach generalizes quite well the Riesz calculus to more general situations and to identify a type of spectral set as the support of the functional calculus. A book on this material is: *Brian Jefferies, Spectral properties of non-commuting operators, Lectures Notes in Mathematics 1843, Springer Verlag, 2004.*

Appendix A

The Orlicz–Pettis Theorem

We establish the Orlicz-Pettis Theorem, which is used when dealing with vector valued measures and integrals. The theorem concerns subseries convergent series in normed spaces.

Definition A.1. If X is a normed space, a formal series $\sum_{j=1}^{\infty} x_j$ in X is weakly subseries convergent if for every subsequence $\{n_j\}$ the subseries $\sum_{j=1}^{\infty} x_{n_j}$ is weakly convergent in X. The remarkable result by Orlicz and Pettis is that any weak subseries convergent series is actually norm subseries convergent ([Or], [Pe]). To establish this result we need a few preliminary results.

Lemma A.2. *If $\{y_j\}$ is norm Cauchy in X and is weakly convergent to $y \in X$, then $\|y_j - y\| \to 0$.*

Proof. Let $\epsilon > 0$ and let
$$S_\epsilon = \{x : \|x\| \leq \epsilon\} = \bigcap_{\|x'\| \leq 1} \{x : |x'(x)| \leq \epsilon\}.$$
Then S_ϵ is a weakly closed norm neighborhood of 0 in X.

There exists N such that $i, j \geq N$ implies $y_i - y_j \in S_\epsilon$. Then $y_i - y_j \to y_i - y$ weakly as $j \to \infty$ so $y_i - y \in S_\epsilon$ for $i \geq N$, since the set S_ϵ is weakly closed. $\qquad \square$

Lemma A.3. *If every series $\sum_j x_j$ in X which is weakly subseries convergent satisfies the condition that $\|x_j\| \to 0$, then every series in X which is weakly subseries convergent is norm subseries convergent.*

Proof. By Lemma A.2 it suffices to show that for every weakly subseries convergent series $\sum_j x_j$, the partial sums $s_n = \sum_{j=1}^{n} x_{n_j}$ of any subseries are norm Cauchy. Suppose such a series exists with partial sums $\{s_n\}$ which

149

are not norm Cauchy. Then there exists $\epsilon > 0$ and an increasing sequence of finite intervals $\{I_k\}$ such that

$$\left\| \sum_{j \in I_k} x_{n_j} \right\| > \epsilon,$$

for every k. But then,

$$\sum_{k=1}^{\infty} \left(\sum_{j \in I_k} x_{n_j} \right) = \sum_{k=1}^{\infty} z_k$$

is a weakly subseries convergent series with $\|z_k\| > \epsilon$, contradicting the hypothesis.   □

Lemma A.4. *Let $x_{ij} \in \mathbb{R}$ and let $\epsilon_{ij} > 0$ for $i, j \in \mathbb{N}$. If $\lim_i x_{ij} = 0$ for every j and $\lim_j x_{ij} = 0$ for every i, then there is a subsequence $\{m_i\}$ such that $|x_{m_i m_j}| < \epsilon_{ij}$ for $i \neq j$.*

Proof.   Put $m_1 = 1$. There exists $m_2 > m_1$ such that

$$|x_{m_1 m}| < \epsilon_{12} \text{ and } |x_{m m_1}| < \epsilon_{21}, \text{ for } m > m_2.$$

There exists $m_3 > m_2$ such that

$$|x_{m_1 m}| < \epsilon_{13}, \ |x_{m_2 m}| < \epsilon_{23}, \ |x_{m m_1}| < \epsilon_{31} \text{ and } |x_{m m_2}| < \epsilon_{32}, \text{ for } m > m_3.$$

Now just continue.   □

We now establish a version of the Antosik-Mikusinski Matrix Theorem which will be used in the proof of the Orlicz-Pettis Theorem. More general versions of the result along with other applications can be found in [Sw3].

Theorem A.5 (Antosik-Mikusinski). *Let $x_{ij} \in X$ for $i, j \in \mathbb{N}$. Suppose*

   (i) *$\lim_i x_{ij} = x_j$ exists for every j and*
   (ii) *for each subsequence $\{m_j\}$ there is a subsequence $\{n_j\}$ of $\{m_j\}$ such that the sequence $\left\{ \sum_{j=1}^{\infty} x_{i n_j} \right\}$ converges.*

Then $\lim_i x_{ij} = x_j$ uniformly for $j \in \mathbb{N}$. Also, $\lim_j x_{ij} = 0$ uniformly for $i \in \mathbb{N}$ and $\lim_i x_{ii} = 0$.

Proof.   If the conclusion fails, there exists $\delta > 0$ and a subsequence $\{k_i\}$ such that

$$\sup_j \|x_{k_i j} - x_j\| > \delta.$$

For notational convenience, assume $k_i = i$. Set $i_1 = 1$ and pick j_1 such that

$$\|x_{i_1 j_1} - x_{j_1}\| > \delta.$$

By (i), there exists $i_2 > i_1$ with

$$\|x_{i_1 j_1} - x_{i_2 j_1}\| > \delta \text{ and } \|x_{ij} - x_j\| < \delta$$

for $i \geq i_2$ and $1 \leq j \leq j_1$. Pick j_2 such that

$$\|x_{i_2 j_2} - x_{j_2}\| > \delta$$

and note $j_2 > j_1$. Continuing by induction produces two increasing sequences $\{i_k\}$ and $\{j_k\}$ such that

$$\left\| x_{i_k j_k} - x_{i_{k+1} j_k} \right\| > \delta.$$

Set $z_{kl} = \left\| x_{i_k j_k} - x_{i_{k+1} j_k} \right\|$ and observe that

$$\|z_{kk}\| > \delta. \tag{A.1}$$

Consider the matrix $M = [z_{kl}]$. By (i), the columns of M converge to 0 and by (ii), the rows of the matrix $[x_{ij}]$ converge to 0 so the same holds for the matrix M. By Lemma A.4 there is a subsequence $\{m_k\}$ such that

$$\|z_{m_k m_l}\| < 1/2^{k+l} \text{ for } k \neq l.$$

By (ii) there is a subsequence $\{n_k\}$ of $\{m_k\}$ such that

$$\lim_k \sum_{l=1}^{\infty} z_{n_k n_l} = 0.$$

Then

$$\|z_{n_k n_k}\| \leq \left\| \sum_{l=1}^{\infty} z_{n_k n_l} \right\| + \left\| \sum_{l \neq k} z_{n_k n_l} \right\| \leq \left\| \sum_{l=1}^{\infty} z_{n_k n_l} \right\| + \sum_{l \neq k} \|z_{n_k n_l}\|$$

$$\leq \left\| \sum_{l=1}^{\infty} z_{n_k n_l} \right\| + \sum_{l=1}^{\infty} 1/2^{l+k} = \left\| \sum_{l=1}^{\infty} z_{n_k n_l} \right\| + 1/2^k. \tag{A.2}$$

Now the first term on the right hand side of (A.2) goes to 0 as $k \to \infty$ by (ii) and since the second term on the right hand side of (A.2) also goes to 0, $\|z_{n_k n_k}\| \to 0$ contradicting (A.1).

The uniform convergence of $\lim_i x_{ij} = x_j$ and the fact that $\lim_j x_{ij} = 0$ for each i implies the double limit, $\lim_{ij} x_{ij}$, exists and equals 0 so, in particular, $\lim_j x_{ij} = 0$ uniformly for $i \in \mathbb{N}$ and $\lim_i x_{ii} = 0$. $\qquad \square$

The Orlicz-Pettis Theorem now follows readily.

Theorem A.6. *(Orlicz-Pettis) If $\sum_j x_j$ is a series in the normed space X which is weakly subseries convergent, then $\sum_j x_j$ is norm subseries convergent.*

Proof. First observe that by replacing X by the span of the $\{x_j\}$, we may assume that X is separable. By Lemma A.3 it suffices to show $\|x_j\| \to 0$.

Pick $x_j' \in X'$ such that

$$|x_j'(x_j)| = \|x_j\|, \|x_j'\| = 1.$$

Since X is separable, the Banach-Alaoglu Theorem implies that there is a subsequence $\{x_{m_j}'\}$ converging weak-$*$ to some $x' \in X'$. Consider the matrix $M = [x_{m_i}'(x_{m_j})]$. We claim the matrix M satisfies the conditions of Theorem A.5. First, condition (i) is satisfied by the weak-$*$ convergence of $\{x_{m_i}'\}$.

If $\{p_j\}$ is any increasing sequence of integers, let $\sum_{j=1}^{\infty} x_{m_{p_j}} = x$ be the weak limit of this subseries. Then

$$\lim_i \sum_{j=1}^{\infty} x_{m_i}'(x_{m_{p_j}}) = \lim_i x_{m_i}'(x)$$

exists, so condition (ii) in Theorem A.5 is also satisfied. By Theorem A.5,

$$x_{m_i}'(x_{m_i}) = \|x_{m_i}\| \to 0.$$

Since the same argument can be applied to any subsequence of $\{x_j\}$, it follows that $\|x_j\| \to 0$. $\square$

Remark A.7. For the history of the Orlicz-Pettis Theorem, see [FL], [Ka] and [DU]. More general forms of the Orlicz-Pettis Theorem for topological vector spaces are presented in [Sw3].

Appendix B

The Spectrum of an Operator

We establish in this appendix some of the basic properties of the spectrum of a continuous linear operator on a Banach space.

Let X be a complex Banach space and let $\mathbf{T} \in L(X)$, the continuous linear operators from X into itself. We first establish a few preliminary results for invertible operators, which are needed in studying the spectrum.

In what follows $\mathcal{R}\mathbf{T}$ will denote the range of the operator $\mathbf{T}$, I will denote the identity operator and $\mathbf{T}^0 = I$.

Theorem B.1. *The operator $\mathbf{T}^{-1}$ exists as a continuous linear operator (on $\mathcal{R}\mathbf{T}$) if and only if there exists $m > 0$ such that $\|\mathbf{T}x\| \geq m\,\|x\|$ for $x \in X$. Moreover, since X is complete $\mathbf{T}$ has closed range.*

Proof.   Suppose $\mathbf{T}^{-1}$ exists, is continuous on $\mathcal{R}\mathbf{T}$ and set $m = 1/\left\|\mathbf{T}^{-1}\right\|$. Then

$$\|x\| = \left\|\mathbf{T}^{-1}(\mathbf{T}x)\right\| \leq \left\|\mathbf{T}^{-1}\right\| \|\mathbf{T}x\|\,.$$

Conversely, assume the inequality holds. Then $\mathbf{T}$ is one-one, so $\mathbf{T}^{-1}$ exists. Moreover,

$$\left\|\mathbf{T}^{-1}(\mathbf{T}x)\right\| = \|x\| \leq \frac{1}{m}\,\|Tx\|$$

which implies that $\mathbf{T}^{-1}$ is continuous on $\mathcal{R}\mathbf{T}$.

For the last statement suppose $\{y_k\} \subset \mathcal{R}\mathbf{T}$ and $y_k \to y$. Let x_k be such that $\mathbf{T}x_k = y_k$. Then

$$\|\mathbf{T}x_k - \mathbf{T}x_j\| \geq m\,\|x_k - x_j\|$$

implies $\{x_k\}$ is Cauchy and converges to some $x \in X$. Hence, $\mathbf{T}x_k = y_k \to y = \mathbf{T}x \in \mathcal{R}\mathbf{T}$. □

Theorem B.2. *If the series $\sum_{j=0}^{\infty} \mathbf{T}^j$ converges in norm in $L(X)$, then $(I - \mathbf{T})$ is invertible and*

$$(I - \mathbf{T})^{-1} = \sum_{j=0}^{\infty} \mathbf{T}^j.$$

Proof.   If $\mathbf{S} = \sum_{j=0}^{\infty} \mathbf{T}^j$, then

$$\mathbf{ST} = \mathbf{TS} = \sum_{j=0}^{\infty} \mathbf{T}^{j+1},$$

so

$$(I - \mathbf{T})\mathbf{S} = \mathbf{S}(I - \mathbf{T}) = I. \qquad \square$$

Corollary B.3. *If $\|\mathbf{T}\| < 1$, then $I - \mathbf{T}$ is invertible and*

$$(I - \mathbf{T})^{-1} = \sum_{j=0}^{\infty} \mathbf{T}^j. \tag{B.1}$$

In this case,

$$\left\|(I - \mathbf{T})^{-1}\right\| \leq \frac{1}{1 - \|\mathbf{T}\|}. \tag{B.2}$$

Proof.   Since $\|\mathbf{T}^j\| \leq \|\mathbf{T}\|^j$, the series $\sum_{j=0}^{\infty} \mathbf{T}^j$ converges absolutely and therefore converges, so the first statement follows from Theorem B.2. For the last statement, we have

$$\left\|(I - \mathbf{T})^{-1}\right\| \leq \sum_{j=0}^{\infty} \|\mathbf{T}\|^j = 1/(1 - \|\mathbf{T}\|).$$

$$\qquad \square$$

The series $\sum_{j=0}^{\infty} \mathbf{T}^j$ is called the *Neumann series* for $(I - \mathbf{T})^{-1}$.

We can improve the sufficient condition $\|\mathbf{T}\| < 1$ for the convergence of the Neumann series. Indeed, we will show that this condition can be replaced by the sufficient condition

$$\lim \sqrt[j]{\|\mathbf{T}^j\|} < 1.$$

We need the following lemma:

Lemma B.4. *Let $a_j \in \mathbb{R}$ satisfy $0 \leq a_{j+k} \leq a_j a_k$ for $j, k \in \mathbb{N}$. Then $\{\sqrt[j]{a_j}\}$ converges to $a = \inf\{\sqrt[j]{a_j} : j \in \mathbb{N}\}$.*

Proof. Let $\epsilon > 0$. Choose j such that $\sqrt[j]{a_j} < a + \epsilon$. For $k > j$, write $k = qj + r$, where $0 \le r \le j - 1$. Set

$$M = \max\{a_r : 0 \le r \le j - 1\}.$$

Then

$$a_k \le a_j ... a_j a_r = (a_j)^q a_r,$$

so

$$a \le \sqrt[k]{a_k} \le \sqrt[k]{a_r}(a_j^{1/j})^{\frac{qj}{qj+r}} \le \sqrt[k]{M}(a + \epsilon)^{\frac{qj}{qj+r}}.$$

Finally,

$$\sqrt[k]{M}(a + \epsilon)^{\frac{qj}{qj+r}} \to a + \epsilon \text{ as } k \to \infty$$

and the result follows. $\square$

Corollary B.5. *For* $\mathbf{T} \in L(X)$ $\lim \sqrt[j]{\|\mathbf{T}^j\|}$ *exists and* $\lim \sqrt[j]{\|\mathbf{T}^j\|} \le \|\mathbf{T}\|$.

Proof. It suffices to set $a_j = \|\mathbf{T}^j\|$ in Lemma B.4. $\square$

The operator $\mathbf{T} : C[0, 1] \to C[0, 1]$ defined by $\mathbf{T}f(t) = \int_0^1 tf(s)ds$ shows that strict inequality can hold in Corollary B.5. Another example is furnished by the Volterra operator which is considered later.

The root test from Calculus ([Ap] 8.26, [DeS] 3.14) carries forward to determine absolute convergence for series of operators.

Theorem B.6. *(Root Test) Let* $\mathbf{T}_j \in L(X)$ *and set*

$$r = \limsup \sqrt[j]{\|\mathbf{T}_j\|}.$$

If $r < 1$, *then the series* $\sum_{j=1}^{\infty} \mathbf{T}_j$ *converges absolutely and if* $r > 1$, *the series* $\sum_{j=1}^{\infty} \mathbf{T}_j$ *diverges (in fact,* $\{\|\mathbf{T}_j\|\}$ *does not converge to zero).*

From Theorems B.2 and B.6, we have a sufficient condition for the convergence of the Neumann series.

Corollary B.7. *If* $\lim \sqrt[j]{\|\mathbf{T}^j\|} < 1$, *the Neumann series* $\sum_{j=0}^{\infty} \mathbf{T}^j$ *converges absolutely in norm to* $(I - \mathbf{T})^{-1}$.

Theorem B.8. *Given* $\mathbf{A} \in L(X)$, *suppose that* $\mathbf{A}^{-1}$ *exists and belongs to* $L(X)$ *as well. If* $\mathbf{B} \in L(X)$ *is such that*

$$\|\mathbf{A} - \mathbf{B}\| < 1/\|\mathbf{A}^{-1}\|,$$

then $\mathbf{B}^{-1} \in L(X)$ *with*

$$\|\mathbf{B}^{-1}\| \le \|\mathbf{A}^{-1}\|/(1 - \|\mathbf{A}^{-1}\|\|\mathbf{A} - \mathbf{B}\|) \tag{B.3}$$

and

$$\|\mathbf{A}^{-1} - \mathbf{B}^{-1}\| \le \|\mathbf{A}^{-1}\|^2 \|\mathbf{A} - \mathbf{B}\|/(1 - \|\mathbf{A}^{-1}\|\|\mathbf{A} - \mathbf{B}\|). \tag{B.4}$$

Proof.   Write
$$B = A - (A - B) = A(I - A^{-1}(A - B)).$$
Since
$$\left\| A^{-1}(A - B) \right\| \leq \left\| A^{-1} \right\| \left\| A - B \right\| < 1,$$
by Corollary B.3,
$$(I - A^{-1}(A - B))^{-1} = \sum_{j=0}^{\infty} [A^{-1}(A - B)]^j.$$
Thus, $B^{-1} \in L(X)$ and
$$B^{-1} = (I - A^{-1}(A - B))^{-1}A^{-1}.$$
Also,
$$B^{-1} = \sum_{j=0}^{\infty} [A^{-1}(A - B)]^j A^{-1},$$
so (B.3) is clear. For (B.4), we have
$$\left\| B^{-1} - A^{-1} \right\| = \left\| (B^{-1}A - I)A^{-1} \right\| \leq \left\| A^{-1} \right\| \left\| B^{-1}A - I \right\|$$
$$= \left\| A^{-1} \right\| \left\| I - \sum_{j=0}^{\infty} [A^{-1}(A - B)]^j \right\|$$
$$= \left\| A^{-1} \right\| \left\| \sum_{j=1}^{\infty} [A^{-1}(A - B)]^j \right\|$$
$$\leq \left\| A^{-1} \right\|^2 \left\| A - B \right\| / (1 - \left\| A^{-1} \right\| \left\| A - B \right\|). \qquad \square$$

Corollary B.9. *The set*
$$\mathcal{G} = \{ B \in L(X) : B^{-1} \in L(X) \}$$
is open in $L(X)$ and the map $B \to B^{-1}$ is continuous from $\mathcal{G}$ onto $\mathcal{G}$.

We now give the definition of resolvent set and of spectrum of an operator.

Definition B.10. A number $\lambda \in \mathbb{C}$ belongs to the resolvent set, $\rho(T)$, of the operator T if $(\lambda - T)^{-1}$ exists and belongs to $L(X)$, where $\lambda - T$ means $\lambda I - T$.

The spectrum of T, $\sigma(T)$, is the complement of $\rho(T)$, $\sigma(T) = \mathbb{C} \backslash \rho(T)$.

A point λ is an eigenvalue of $\mathbf{T}$ with eigenvector $x \neq 0$ if $(\lambda - \mathbf{T})x = 0$. Note that any eigenvalue belongs to $\sigma(\mathbf{T})$. The set of all eigenvalues is called the point spectrum of $\mathbf{T}$ and is denoted by $P\sigma(\mathbf{T})$.

The points of the spectrum are further classified. All points $\lambda \in \sigma(\mathbf{T})$ such that $\lambda - \mathbf{T}$ is one-one and $\mathcal{R}(\lambda - \mathbf{T})$ is dense in X is called the continuous spectrum and is denoted by $C\sigma(\mathbf{T})$. All points $\lambda \in \sigma(\mathbf{T})$ such that $\lambda - \mathbf{T}$ is one-one and $\mathcal{R}(\lambda - \mathbf{T})$ is not dense in X is called the residual spectrum and is denoted by $R\sigma(\mathbf{T})$.

Thus, the spectrum of $\mathbf{T}$ is the disjoint union of $P\sigma(\mathbf{T}), C\sigma(\mathbf{T})$ and $R\sigma(\mathbf{T})$.

If X is finite dimensional, then the spectrum of an operator $\mathbf{T}$ is just the set of eigenvalues of $\mathbf{T}$.

Theorem B.11. *The resolvent set $\rho(\mathbf{T})$ is open and the spectrum $\sigma(\mathbf{T})$ is closed.*

Proof. Suppose $\lambda \in \rho(\mathbf{T})$ and put

$$M = \left\| (\lambda - \mathbf{T})^{-1} \right\|.$$

If $|\lambda - \mu| < 1/M$, then

$$(\mu - \mathbf{T}) - (\lambda - \mathbf{T}) = (\mu - \lambda)I$$

so by Theorem B.8, $\mu \in \rho(T)$. $\qquad\qquad\qquad\qquad\qquad\qquad\square$

For $\lambda \in \rho(\mathbf{T})$ we write

$$R_\lambda = R_\lambda(T) = (\lambda - \mathbf{T})^{-1}.$$

The operator R_λ is called the *resolvent operator* of $\mathbf{T}$.

Theorem B.12. *Suppose $\lambda, \mu \in \rho(\mathbf{T})$. Then*

(i) $R_\lambda - R_\mu = (\mu - \lambda)R_\lambda R_\mu$ *and*
(ii) $R_\lambda R_\mu = R_\mu R_\lambda$.
(iii) *If the operator $\mathbf{S} \in L(X)$ commutes with $\mathbf{T}$, then $\mathbf{S}$ commutes with R_λ.*

Proof. Observe that

$$(\mu - \mathbf{T})(\lambda - \mathbf{T})(R_\lambda - R_\mu) = \mu - \mathbf{T} - (\lambda - \mathbf{T}) = (\mu - \lambda)I.$$

Multiplying this equation by $R_\lambda R_\mu$ gives (i). By symmetry

$$R_\mu - R_\lambda = (\lambda - \mu)R_\mu R_\lambda$$

and adding this to (i) gives (ii).

For (iii), using commutativity we have

$$R_\lambda S(\lambda - \mathbf{T}) = R_\lambda(\lambda - \mathbf{T})S = S = SR_\lambda(\lambda - \mathbf{T})$$

which implies $R_\lambda S = SR_\lambda$.   □

The equation

$$R_\lambda - R_\mu = (\mu - \lambda)R_\lambda R_\mu$$

is called the *resolvent equation*.

Theorem B.13. *If* $|\lambda| > \lim \sqrt[j]{\|\mathbf{T}^j\|}$ *(recall this limit exists), then*

$$(\lambda - \mathbf{T})^{-1} \in L(X)$$

and

$$R_\lambda = \sum_{j=1}^{\infty}(1/\lambda^j)\mathbf{T}^{j-1},$$

where this series is norm convergent in $L(X)$ *and*

$$\|R_\lambda\| \leq 1/(|\lambda| - \|\mathbf{T}\|).$$

Moreover, if the series $\sum_{j=1}^{\infty}(1/\lambda^j)\mathbf{T}^{j-1}$ *converges for some value* λ, *then* $\lambda \in \rho(\mathbf{T})$ *and the series equals* R_λ.

Proof.   Since $\lambda - \mathbf{T} = \lambda(I - \mathbf{T}/\lambda)$ and

$$\lim \sqrt[n]{\|(\mathbf{T}/\lambda)^n\|} = (1/|\lambda|)\lim \sqrt[n]{\|\mathbf{T}^n\|} < 1,$$

it follows from Corollary B.7 that the Neumann series for $(I - \mathbf{T}/\lambda)^{-1}$,

$$(I - \mathbf{T}/\lambda)^{-1} = \sum_{n=0}^{\infty}\mathbf{T}^n/\lambda^n,$$

converges in $L(X)$. So,

$$(\lambda - \mathbf{T})^{-1} = \sum_{n=1}^{\infty}(1/\lambda^n)\mathbf{T}^{n-1}.$$

For the last inequality,

$$\left\|(\lambda - \mathbf{T})^{-1}\right\| = \left\|\sum_{n=1}^{\infty}(1/\lambda^n)\mathbf{T}^{n-1}\right\| \leq (1/|\lambda|)\sum_{n=0}^{\infty}\|\mathbf{T}/\lambda\|^n = 1/(|\lambda| - \|\mathbf{T}\|).$$

The last statement follows from the fact that $(\lambda - \mathbf{T})\sum_{j=1}^{\infty}(1/\lambda^j)\mathbf{T}^{j-1} = (\sum_{j=1}^{\infty}(1/\lambda^j)\mathbf{T}^{j-1})(\lambda - \mathbf{T}) = I.$ □

Corollary B.14. *The spectrum $\sigma(T)$ is compact with*

$$\sigma(\mathbf{T}) \subset \{\lambda \in \mathbb{C} : |\lambda| \le \lim \sqrt[n]{\|(\mathbf{T}/\lambda)^n\|}\} .$$

Proof. $\sigma(\mathbf{T})$ is closed by Theorem B.11 and is bounded by $\lim \sqrt[n]{\|\mathbf{T}^n\|}$ from Theorem B.13. □

We next show the spectrum is non-empty.

Theorem B.15. $\sigma(\mathbf{T}) \ne \emptyset$.

Proof. Let $z \in L(X)'$. Define $f = f_z : \rho(\mathbf{T}) \to \mathbb{C}$ by

$$f(\lambda) = z(R_\lambda) = z((\lambda - \mathbf{T})^{-1}).$$

By the resolvent equation

$$(f(\lambda) - f(\mu))/(\lambda - \mu) = z((R_\lambda - R_\mu)/(\lambda - \mu)) = z(-R_\mu R_\lambda).$$

By Corollary B.9 the map $\lambda \to R_\lambda$ from $\rho(\mathbf{T})$ to $L(X)$ is continuous so letting $\lambda \to \mu$ gives

$$\lim_{\lambda \to \mu} (f(\lambda) - f(\mu))/(\lambda - \mu) = -z(R_\mu^2).$$

Thus, f is analytic on $\rho(\mathbf{T})$ with $f'(\mu) = -z(R_\mu^2)$.

Moreover, f is bounded for large λ since

$$|f(\lambda)| \le \|z\| \, \|R_\lambda\| \le \|z\| / (|\lambda| - \|\mathbf{T}\|)$$

for $|\lambda| > \|\mathbf{T}\|$ by Theorem B.13. Hence, if $\sigma(\mathbf{T}) = \emptyset$, then f would be a bounded entire function and by Liouville's Theorem, f would be constant. Since $f(\lambda) \to 0$ as $|\lambda| \to \infty$, f would have to be 0. Since $z \in L(X)'$ is arbitrary, this would imply that $R_\lambda = 0$ which is not possible since R_λ is an inverse. □

Corollary B.16. *If $z \in L(X)'$, then the function $f(\lambda) = z(R_\lambda)$ is analytic on $\rho(\mathbf{T})$ with $f'(\lambda) = -z(R_\lambda^2)$.*

In Chapter 7 we will discuss vector valued analytic functions. In the language of Chapter 7 we would say that R_λ is weakly analytic. It follows easily from the resolvent equation that R_λ is strongly analytic in the sense that

$$\lim_{\lambda \to \mu} (R_\lambda - R_\mu)/(\lambda - \mu) = -R_\mu^2 = \frac{d}{d\mu} R_\mu$$

in norm. This follows from the proof above.

In this appendix we have been dealing with continuous linear operators defined on a Banach space. When dealing with linear operators which are not necessarily continuous and defined on normed spaces which may not be complete the situation is completely different. The spectrum may be empty or be all of $\mathbb{C}$. See [Sw2] for examples.

From Theorem B.15 the spectrum of a continuous linear operator is a non-empty compact subset of $\mathbb{C}$. The following example shows that any compact subset of $\mathbb{C}$ is the spectrum of a continuous linear operator on l^2.

Example B.17. Let K be any compact subset of $\mathbb{C}$, $\{d_j : j \in \mathbb{N}\}$ a dense subset of K and $M = \max\{|z| : z \in K\}$. Define a linear operator $\mathbf{D} : l^2 \to l^2$ by $\mathbf{D}\{t_j\} = \{d_j t_j\}$ where $\{t_j\} \in l^2$; that is, $\mathbf{D}$ is the matrix diagonal map with the $\{d_j\}$ down the diagonal. Note $\mathbf{D}$ is well defined, has values in l^2 and is continuous;

$$\|\mathbf{D}\{t_j\}\|_2^2 = \sum_{j=1}^{\infty} |d_j t_j|^2 \leq M^2 \|\{t_j\}\|_2^2.$$

We claim the spectrum of $\mathbf{D}$ is exactly K. If $\lambda \in \mathbb{C} \setminus K$, there exists $d > 0$ such that $|\lambda - d_j| > d$ for every j. Then

$$\|(\lambda - \mathbf{D})\{y_j\}\|_2^2 = \sum_{j=1}^{\infty} |(\lambda - d_j)t_j|^2 \geq d^2 \|\{t_j\}\|_2^2$$

so $\lambda - \mathbf{D}$ has a continuous inverse and since $\lambda - D$ is onto, $\lambda \in \rho(\mathbf{D})$. Hence, $\sigma(\mathbf{D}) \subset K$. Since $\mathbf{D}e^j = d_j e^j$, each d_j is an eigenvalue with associated eigenvector d_j. Hence, $K = \overline{\{d_j : j \in \mathbb{N}\}} = \sigma(\mathbf{D})$ since the spectrum is closed. Note that the same computations hold for any l^p, $1 \leq p < \infty$. Also, if $d_j \to 0$, then $\mathbf{D}$ is an example of a compact operator with spectrum $\{d_j : j \in \mathbb{N}\} \cup \{0\}$.

Definition B.18. The spectral radius of $\mathbf{T}, r(\mathbf{T})$, is defined to be
$$r(\mathbf{T}) = \sup\{|\lambda| : \lambda \in \sigma(\mathbf{T})\}.$$

Theorem B.19. $r(\mathbf{T}) \leq \|\mathbf{T}^j\|^{1/j}$ *for every* $j \geq 1$ *and* $r(\mathbf{T}) = \lim \sqrt[j]{\|\mathbf{T}^j\|}$.

Proof. From Corollary B.14, we have $r(\mathbf{T}) \leq \lim \sqrt[n]{\|\mathbf{T}^n\|}$. The series
$$\sum_{n=0}^{\infty} (1/\lambda^{n+1})\mathbf{T}^n$$
converges in $L(X)$ to R_λ for $|\lambda| > \|\mathbf{T}\|$. From Corollary B.16, for every $z \in L(X)'$, the series $\sum_{n=0}^{\infty} (1/\lambda^{n+1})z(\mathbf{T}^n)$ converges to the analytic function $-z(R_\lambda^2)$ for $|\lambda| > r(\mathbf{T})$. Therefore, for $|\lambda| > r(\mathbf{T})$,
$$\sup\{|z(\mathbf{T}^n)/\lambda^{n+1}| : n\} < \infty$$

so by the Uniform Boundedness Principle,

$$\sup\{\|\mathbf{T}^n/\lambda^{n+1}\| : n\} = M < \infty.$$

Hence, $\|\mathbf{T}^n\| \le M\,|\lambda^{n+1}|$ and

$$\lim \sqrt[n]{\|\mathbf{T}^n\|} \le \lim(M\,|\lambda|^{n+1})^{1/n} = |\lambda|$$

for $|\lambda| > r(\mathbf{T})$ so $\lim \sqrt[n]{\|\mathbf{T}^n\|} \le r(\mathbf{T})$. $\qquad\qquad\square$

We will give an example to show that strict inequality can hold in the theorem above. The example is the Volterra operator. The computations concerning the operator give interesting examples of the series for the resolvent of the operator and the resolvent.

Example B.20. Let $X = C[0,1]$ with the sup norm,

$$\|f\| = \max\{|f(t)| : 0 \le t \le 1\}.$$

Define $\mathbf{V} : X \to X$ by

$$\mathbf{V}f(t) = \int_0^t f(s)ds.$$

Obviously $\mathbf{V}$ is linear and $\mathbf{V}$ is continuous since $\|\mathbf{V}f\| \le \|f\|$ for $f \in X$. Thus, $\|\mathbf{V}\| \le 1$ and $\|\mathbf{V}\| = 1$ since $\|\mathbf{V}(1)\| = 1$. It is easily shown by induction that

$$\mathbf{V}^{n+1}f(t) = (1/n!)\int_0^t (t-s)^n f(s)ds$$

and then

$$\|\mathbf{V}^{n+1}f\| \le (1/(n+1)!)\,\|f\|.$$

This means that if $\lambda \ne 0$ the series for the resolvent $\sum_{n=1}^{\infty} \lambda^{-n}V^{n-1}$ converges absolutely to R_λ and $\lambda \in \rho(\mathbf{V})$ (Theorem B.13). That is, $\sigma(\mathbf{V}) = \{0\}$. Thus $r(\mathbf{V}) = 0 < \|\mathbf{V}\| = 1$ and strict inequality in the theorem above can hold. Moreover we can use the series for the resolvent to compute a useful formula for the resolvent. We have

$$R_\lambda f(t) = \sum_{n=1}^{\infty} \lambda^{-n} \mathbf{V}^{n-1} f(t)$$

$$= \frac{1}{\lambda} f(t) + \sum_{n=2}^{\infty} \lambda^{-n} \int_0^t \frac{1}{(n-2)!} (t-s)^{n-2} f(s) ds$$

$$= \frac{1}{\lambda} f(t) + \frac{1}{\lambda^2} \int_0^t \sum_{n=2}^{\infty} ((\frac{t-s}{\lambda})^{n-2}/(n-2)!) f(s) ds$$

$$= \frac{1}{\lambda} f(t) + \frac{1}{\lambda^2} \int_0^t e^{(t-s)/\lambda} f(s) ds.$$

We give two further examples of operators and their spectrums.

Example B.21. Let $X = l^p, 1 \leq p < \infty$, and let $\mathbf{L}$ be the left shift, $\mathbf{L}\{t_1, t_2, ...\} = \{t_2, t_3, ...\}$, $\{t_j\} \in l^p$. Since $\|\mathbf{L}\| = 1$, $\sigma(\mathbf{L}) \subset \{\lambda : |\lambda| \leq 1\}$. If $|\lambda| < 1$, then λ is an eigenvalue with associated eigenvector $\{1, \lambda, \lambda^2, ...\}$ so $\sigma(\mathbf{L}) = \{\lambda : |\lambda| \leq 1\}$. Note if $|\lambda| = 1$, then λ is not an eigenvalue since $\{1, \lambda, \lambda^2, ...\} \notin l^p$.

Example B.22. Let $X = C[0,1]$ and define $\mathbf{T} : X \to X$ by $\mathbf{T}f(t) = tf(t)$. Then $\|\mathbf{T}\| = 1$ so $\sigma(\mathbf{T}) \subset \{\lambda : |\lambda| \leq 1\}$. Suppose $\lambda = a + bi$ with $|\lambda| \leq 1$ and $b \neq 0$. Then

$$|(\lambda - \mathbf{T})f(t)|^2 = |(a-t)f(t) + bf(t)i|^2$$
$$= (a-t)^2 |f(t)|^2 + b^2 |f(t)|^2 \geq b^2 |f(t)|^2$$

and

$$\|(\lambda - \mathbf{T})f\| \geq |b| \|f\|$$

so $\lambda - \mathbf{T}$ has a continuous inverse and $\lambda \in \rho(\mathbf{T})$ since $\lambda - \mathbf{T}$ is clearly onto X. Hence, $\sigma(\mathbf{T}) \subset [-1, 1]$. Similarly, if $-1 \leq \lambda < 1$,

$$|(\lambda - \mathbf{T})f(t)| = |(\lambda - t)| |f(t)| \geq |\lambda| |f(t)|$$

which implies

$$\|(\lambda - \mathbf{T})f\| \geq |\lambda| \|f\|$$

so $\lambda - \mathbf{T}$ has a continuous inverse and is onto so $\lambda \in \rho(\mathbf{T})$. Hence, $\sigma(\mathbf{T}) \subset [0, 1]$. Let $\lambda \in [0, 1]$ and $\epsilon > 0$ be such that $[\lambda, \lambda + \epsilon]$ or $[\lambda - \epsilon, \lambda]$ is contained in $[0, 1]$. Assume the former. Construct f_ϵ to be 1 on $[\lambda + \epsilon/3, \lambda + \epsilon]$, 0 on $[0, \lambda]$ and $[\lambda + \epsilon, 1]$ and linear on $[\lambda, \lambda + \epsilon/3] \cup [\lambda + 2\epsilon/3, \lambda + \epsilon]$. Then $\|f_\epsilon\| = 1$ and

$$\|(\lambda - \mathbf{T})f_\epsilon\| \leq \epsilon$$

so

$$\|(\lambda - \mathbf{T})f_\epsilon\| \to 0$$

as $\epsilon \to 0$. So $\lambda - \mathbf{T}$ does not have a continuous inverse (B.1) and $\lambda \in \sigma(\mathbf{T})$. Hence, $\sigma(\mathbf{T}) = [0, 1]$. Note that no point $\lambda \in [0, 1]$ is a eigenvalue of $\mathbf{T}$ $[(\lambda - \mathbf{T})f(t) = 0$ implies $f = 0$ except possibly at $t = \lambda$ so $f = 0]$. Note also the resolvent operator R_λ is given by $R_\lambda f(t) = (1/(\lambda - t))f(t)$, $\lambda \notin [0, 1]$, $f \in X$.

A point $\lambda \in \mathbb{C}$ is called an approximate eigenvalue of $\mathbf{T}$ if to each $\epsilon > 0$ there correspond $x = x_\epsilon$ such that $\|x_\epsilon\| = 1$ and $\|(\lambda - \mathbf{T})x\| \leq \epsilon$. Note an approximate eigenvalue must belong to $\sigma(\mathbf{T})$. The set of all approximate eigenvalues is called the approximate spectrum of $\mathbf{T}$. In the example above we have shown that every point in the spectrum of $\mathbf{T}$ is an approximate eigenvalue. In Corollary C.15 it is shown that for a normal operator its spectrum coincides with its approximate spectrum.

Theorem B.23. *(Spectral Mapping Theorem: Junior Grade) If p is a complex polynomial, then $\sigma(p(\mathbf{T})) = p(\sigma(\mathbf{T}))$.*

Proof. Fix $\mu \in \mathbb{C}$ and let $p(z) - \mu = c \prod_{j=1}^{n} (z - b_j)$ so

$$p(\mathbf{T}) - \mu = c \prod_{j=1}^{n} (\mathbf{T} - b_j). \tag{B.5}$$

If $\mu \in \sigma(\mathbf{T})$, then by (B.5) some $b_j \in \sigma(\mathbf{T})$ since otherwise $p(\mathbf{T}) - \mu$ would be invertible. But, $p(b_j) = \mu$ so $\mu \in p(\sigma(\mathbf{T}))$ and $\sigma(p(\mathbf{T})) \subset p(\sigma(\mathbf{T}))$.

Suppose some $b_j \in \sigma(\mathbf{T})$, say, b_1. If $\mathbf{T} - b_1$ has a bounded inverse, then the range of $\mathbf{T} - b_1$ is not X and (B.5) implies the range of $p(\mathbf{T}) - \mu$ is not X. Then $\mu \in \sigma(p(\mathbf{T}))$. If $\mathbf{T} - b_1$ doesn't have a bounded inverse, then (B.5) implies that $p(\mathbf{T}) - \mu$ has no inverse so in this case $\mu \in \sigma(p(\mathbf{T}))$. Hence, $\sigma(p(\mathbf{T})) \supset p(\sigma(\mathbf{T}))$. $\square$

We also have the relationship between the spectrum of an operator $\mathbf{T}$ and its adjoint or transpose $\mathbf{T}'$. The relationship will follow immediately from the following lemma.

Lemma B.24. T $\in L(X)$ *is invertible iff* **T**′ *is invertible. In this case* $(\mathbf{T}')^{-1} = (\mathbf{T}^{-1})'$.

Proof. Suppose **T** is invertible. Then $\mathbf{TT}^{-1} = \mathbf{T}^{-1}\mathbf{T} = I$ implies $\mathbf{T}'(\mathbf{T}^{-1})' = (\mathbf{T}^{-1})'\mathbf{T}' = I$ so **T**′ is invertible and $(\mathbf{T}')^{-1} = (\mathbf{T}^{-1})'$.

For the other implication by what has just been established **T**″ is invertible. Now **T**″ restricted to X is **T** [with X identified with its image in X'' under the cannonical imbedding] so **T** is one-one and has closed range since **T** has a continuous inverse and X is complete (B.1).

It suffices to show $\mathcal{R}\mathbf{T} = X$. Suppose $\mathcal{R}\mathbf{T} \subsetneq X$. By a corollary of the Hahn-Banach Theorem there exists $x' \in X', x' \neq 0$, such that $x'(\mathbf{T}X) = \{0\}$. But then $\mathbf{T}'x'(x) = 0$ for all $x \in X$ so $\mathbf{T}'x' = 0$. But, this means **T**′ is not one-one.   □

Theorem B.25. $\sigma(\mathbf{T}) = \sigma(\mathbf{T}')$ *and* $(\lambda - \mathbf{T}')^{-1} = ((\lambda - \mathbf{T})^{-1})'$ *or* $R_\lambda(\mathbf{T}') = R_\lambda(\mathbf{T})'$.

Appendix C

Self Adjoint, Normal and Unitary Operators

In this appendix we define and describe some of the basic properties of self adjoint, normal and unitary operators on a Hilbert space. Let H be a complex Hilbert space and denote the inner or dot product on H by $x \cdot y$.

We first recall the definition of the Hilbert space adjoint which is somewhat different from the transpose or adjoint operator between normed or locally convex spaces. If $y \in H$, then the map $f_y : H \to \mathbb{C}$ defined by $f_y(x) = x \cdot y$ gives a continuous linear functional on H with $\|f\| = \|f_y\|$. The Riesz Representation Theorem for Hilbert space asserts that every continuous linear functional on H is of this form ([Sw2] Appendix, Theorem 19). The map $\Psi : H \to H'$ given by $\Psi y = f_y$ is an additive isometry from H onto H' but is only conjugate homogeneous (that is, $\Psi(tx) = \bar{t}x$). If we identify H and H' under the map Ψ, we define the Hilbert space adjoint of $\mathbf{T}$, denote by $\mathbf{T}^*$, by $\mathbf{T}^* = \Psi^{-1}\mathbf{T}'\Psi$, where $\mathbf{T}'$ is the usual adjoint or transpose operator. Thus, if $x, y \in H$, the adjoint operator is characterized by the condition

$$\mathbf{T}y \cdot x = y \cdot \mathbf{T}^*x,$$

since $\mathbf{T}'\Psi x(y) = \Psi(x)(\mathbf{T}y) = \mathbf{T}y \cdot x$ and $\mathbf{T}'\Psi x(y) = \Psi\mathbf{T}^*x(y) = y \cdot \mathbf{T}x$.

The Hilbert space adjoint has the following properties:

$$(\mathbf{T} + \mathbf{S})^* = \mathbf{T}^* + \mathbf{S}^*, (\lambda\mathbf{T})^* = \bar{\lambda}\mathbf{T}^*, (\mathbf{TS})^* = \mathbf{S}^*\mathbf{T}^*,$$

$$\mathbf{T}^{**} = \mathbf{T}, I^* = I, \|\mathbf{T}\| = \|\mathbf{T}^*\|$$

and if either $\mathbf{T}^{-1}$ or $(\mathbf{T}^*)^{-1}$ exists and is in $L(H)$, the other exists and

$$(\mathbf{T}^{-1})^* = (\mathbf{T}^*)^{-1}.$$

It follows from these properties that the spectrums of $\mathbf{T}$ and $\mathbf{T}^*$ are related by the equality

$$\sigma(\mathbf{T}^*) = \overline{\sigma(\mathbf{T})}.$$

For example, if $\{d_j\} \subset \mathbb{C}$ is bounded and $\mathbf{D} : l^2 \to l^2$ is the diagonal operator defined by $\mathbf{D}\{t_j\} = \{d_j t_j\}$, then $\mathbf{D}^*$ is the diagonal operator $\mathbf{D}^* : l^2 \to l^2$, $\mathbf{D}^*\{t_j\} = \{\bar{d}_j t_j\}$.

Remark C.1. Since $\|\mathbf{T}\| = \|\mathbf{T}^*\|$ the map $T \to \mathbf{T}^*$ from $L(H) \to L(H)$ is continuous.

A continuous linear operator $\mathbf{T} \in L(H)$ is *self adjoint* if $\mathbf{T} = \mathbf{T}^*$ where $\mathbf{T}^*$ is the Hilbert space adjoint of $\mathbf{T}$. Note that if $\mathbf{T}$ is self adjoint, then

$$\mathbf{T}x \cdot y = x \cdot \mathbf{T}^* y = x \cdot \mathbf{T}y$$

so $\mathbf{T}$ is symmetric.

Proposition C.2. *The self adjoint operators in $L(H)$ form a (norm) closed real linear subspace containing I.*

Proof.   If $\mathbf{T}, \mathbf{S} \in L(H)$ are self adjoint and $t, s \in \mathbb{R}$, then

$$(t\mathbf{T} + s\mathbf{S})^* = t\mathbf{T}^* + s\mathbf{S}^* = t\mathbf{T} + s\mathbf{S}$$

so the self adjoint operators form a real linear subspace.
    If $\{\mathbf{T}_j\}$ are self adjoint, $\mathbf{T} \in L(H)$ and $\|\mathbf{T}_j - \mathbf{T}\| \to 0$, then

$$\|\mathbf{T} - \mathbf{T}^*\| \leq \|\mathbf{T} - \mathbf{T}_j\| + \|\mathbf{T}_j^* - \mathbf{T}^*\| \to 0$$

so $\mathbf{T} = \mathbf{T}^*$.   □

Proposition C.3. *If $\mathbf{T}, \mathbf{S} \in L(H)$ are self adjoint, then $\mathbf{TS}$ is self adjoint iff $\mathbf{TS} = \mathbf{ST}$.*

Proof. $\Longrightarrow$: $(\mathbf{TS})^* = \mathbf{S}^*\mathbf{T}^* = \mathbf{ST} = \mathbf{TS}$.
    $\Longleftarrow$: $(\mathbf{ST})^* = \mathbf{T}^*\mathbf{S}^* = \mathbf{TS} = \mathbf{ST}$.   □

Lemma C.4. *Let X be a complex inner product space and $\mathbf{T} \in L(X)$. If $\mathbf{T}x \cdot x = 0$ for every $x \in X$, then $\mathbf{T} = 0$.*

Proof.   Let $x, y \in X, s, t \in \mathbb{C}$. Then

$$0 = \mathbf{T}(sx + ty) \cdot (sx + ty) = |s|^2\, \mathbf{T}x \cdot x + |t|^2\, \mathbf{T}y \cdot y + s\bar{t}\mathbf{T}x \cdot y + \bar{s}t\mathbf{T}y \cdot x$$

so

$$0 = s\bar{t}\mathbf{T}x \cdot y + \bar{s}t\mathbf{T}y \cdot x . \tag{C.1}$$

Put $s, t = 1$ in (C.1) to obtain $\mathbf{T}x \cdot y + \mathbf{T}y \cdot x = 0$; put $s = i, t = 1$ in (C.1) to obtain $i\mathbf{T}x \cdot y - i\mathbf{T}y \cdot x = 0$. Thus, $2\mathbf{T}x \cdot y = 0$ and $\mathbf{T} = 0$.   □

The use of complex scalars in Lemma C.4 is important; consider rotations in the plane.

Theorem C.5. T $\in L(H)$ *is self adjoint iff* $\mathbf{T}x \cdot x \in \mathbb{R}$ *for every* $x \in H$.

Proof. If **T** is self adjoint, then

$$\mathbf{T}x \cdot x = x \cdot \mathbf{T}x = \overline{\mathbf{T}x \cdot x}$$

so $\mathbf{T}x \cdot x$ is real.

If $\mathbf{T}x \cdot x \in \mathbb{R}$ for every $x \in H$, then

$$\mathbf{T}x \cdot x = \overline{\mathbf{T}x \cdot x} = \overline{x \cdot \mathbf{T}^*x} = \mathbf{T}^*x \cdot x$$

so

$$(\mathbf{T} - \mathbf{T}^*)x \cdot x = 0.$$

Hence, $\mathbf{T} - \mathbf{T}^* = 0$ by Lemma C.4. $\qquad\square$

In the algebra $L(H)$ the self adjoint operators perform a role analogous to the real numbers in the complex numbers as the following result suggests.

Theorem C.6. *If* $\mathbf{T} \in L(H)$, *then there exist unique self adjoint operators* **A** *and* **B** *such that*

$$\mathbf{T} = \mathbf{A} + i\mathbf{B}.$$

Proof. Set $\mathbf{A} = (\mathbf{T} + \mathbf{T}^*)/2$ and $\mathbf{B} = (\mathbf{T} - \mathbf{T}^*)/2i$.

For uniqueness, suppose $\mathbf{T} = \mathbf{A}_1 + i\mathbf{B}_1$ with $\mathbf{A}_1, \mathbf{B}_1$ self adjoint. If $x \in H$, then

$$\mathbf{A}x \cdot x + i\mathbf{B}x \cdot x = \mathbf{A}_1 x \cdot x + i\mathbf{B}_1 x \cdot x$$

and by Theorem C.5

$$\mathbf{A}x \cdot x = \mathbf{A}_1 x \cdot x, \mathbf{B}x \cdot x = \mathbf{B}_1 x \cdot x$$

so $\mathbf{A} = \mathbf{A}_1, \mathbf{B} = \mathbf{B}_1$ by Lemma C.4. $\qquad\square$

We can also define an order on the self adjoint operators.

Definition C.7. Let $\mathbf{T}, \mathbf{S} \in L(H)$ be self adjoint. Then $\mathbf{T} \geq \mathbf{S}$ iff

$$\mathbf{T}x \cdot x \geq \mathbf{S}x \cdot x$$

for every $x \in H$. If $T \geq 0$, we say that **T** is a positive operator. Note this definition is meaningful by Theorem C.5.

If $\{d_j\} \subset \mathbb{R}$ is bounded and $\mathbf{D} : l^2 \to l^2$ is the diagonal operator $\mathbf{D}\{t_j\} = \{d_j t_j\}$, then $\mathbf{D}$ is self adjoint and is positive iff $d_j \geq 0$ for all j.

Proposition C.8. *Let $\mathbf{U}, \mathbf{T}, \mathbf{S} \in L(H)$ be self adjoint. If $\mathbf{T} \geq \mathbf{S}$, then $\mathbf{T} + \mathbf{U} \geq \mathbf{S} + \mathbf{U}$ and $t\mathbf{T} \geq t\mathbf{S}$ for $t \geq 0$. This order is a partial order on the self adjoint operators.*

Proof. The first statement is clear. For the last statement, clearly $\mathbf{T} \geq \mathbf{T}$ and if $\mathbf{T} \geq \mathbf{S}$ and $\mathbf{S} \geq \mathbf{U}$, then $\mathbf{T} \geq \mathbf{U}$. If $\mathbf{T} \geq \mathbf{S}$ and $\mathbf{S} \geq \mathbf{T}$, then $(\mathbf{T} - \mathbf{S})x \cdot x = 0$ for every x so $\mathbf{T} = \mathbf{S}$ by Lemma C.4. $\qquad\square$

In the real numbers any positive number has a unique square root; we will establish the analogue of this statement in the chapter on self adjoint operators. In Theorem C.23 we will also establish a sequential completeness result analogous to that for the real numbers.

Recall that if $M \subset H$, then

$$M^\perp = \{x : x \cdot m = 0 \text{ for every } m \in M\}.$$

Proposition C.9. *If $\mathbf{T} \in L(H)$, then $(\overline{\mathcal{R}\mathbf{T}})^\perp = \ker \mathbf{T}^*$ and $\overline{\mathcal{R}\mathbf{T}} = (\ker \mathbf{T}^*)^\perp$.*

Proof. Let $y \in \ker \mathbf{T}^*$ so $\mathbf{T}^* y = 0$ and

$$\mathbf{T}^* y \cdot x = y \cdot \mathbf{T}x = 0$$

for all $x \in H$. Hence, $y \in \mathcal{R}\mathbf{T}^\perp = (\overline{\mathcal{R}\mathbf{T}})^\perp$.
  Let $y \in \mathcal{R}\mathbf{T}^\perp = \overline{\mathcal{R}\mathbf{T}}^\perp$. Then

$$y \cdot \mathbf{T}x = \mathbf{T}^* y \cdot x = 0$$

so $y \in \ker \mathbf{T}^*$ and $\mathcal{R}\mathbf{T}^\perp = \overline{\mathcal{R}\mathbf{T}}^\perp \subset \ker \mathbf{T}^*$.
  Hence, $(\overline{\mathcal{R}\mathbf{T}})^\perp = \ker \mathbf{T}^*$ and the second identity follows. $\qquad\square$

Definition C.10. An operator $\mathbf{T} \in L(H)$ is normal if

$$\mathbf{T}\mathbf{T}^* = \mathbf{T}^*\mathbf{T}.$$

A self adjoint operator is obviously normal but not conversely. If $\mathbf{D} : l^2 \to l^2$ is the diagonal operator $\mathbf{D}\{t_j\} = \{d_j t_j\}$ ($\{d_j\}$ bounded in $\mathbb{C}$), then $\mathbf{D}^*$ is the diagonal operator $\mathbf{D}^*\{t_j\} = \{\overline{d_j} t_j\}$ so $\mathbf{D}$ is normal but is only self adjoint if $\{d_j\} \subset \mathbb{R}$. If an operator is normal we show that its range and kernel give an orthogonal decomposition for H.

Theorem C.11. *Let* $\mathbf{T} \in L(H)$. *Then*

$$\|\mathbf{TT}^*\| = \|\mathbf{T}^*\mathbf{T}\| = \|\mathbf{T}\|^2$$

and $\mathbf{T}$ *is normal iff*

$$\|\mathbf{T}^*x\| = \|\mathbf{T}x\|$$

for every $x \in H$.

Proof. First,

$$
\begin{aligned}
\|\mathbf{TT}^*\| &\leq \|\mathbf{T}\|\,\|\mathbf{T}^*\| \\
&= \|\mathbf{T}\|^2 \\
&= (\sup\{\|\mathbf{T}x\| : \|x\| \leq 1\})^2 \\
&= \sup\{\|\mathbf{T}x\|^2 : \|x\| \leq 1\} \\
&= \sup\{\mathbf{T}x \cdot \mathbf{T}x : \|x\| \leq 1\} \\
&= \sup\{\mathbf{T}^*\mathbf{T}x \cdot x : \|x\| \leq 1\} \\
&\leq \sup\{\|\mathbf{T}^*\mathbf{T}x\| : \|x\| \leq 1\} = \|\mathbf{T}^*\mathbf{T}\|.
\end{aligned}
$$

For $x \in H$, we have

$$\|\mathbf{T}x\|^2 = \mathbf{T}x \cdot \mathbf{T}x = \mathbf{T}^*\mathbf{T}x \cdot x$$

and

$$\|\mathbf{T}^*x\|^2 = \mathbf{T}^*x \cdot \mathbf{T}^*x = \mathbf{TT}^*x \cdot x.$$

If $\mathbf{T}$ is normal, these equations imply $\|\mathbf{T}x\|^2 = \|\mathbf{T}^*x\|^2$. On the other hand, if $\|\mathbf{T}^*x\| = \|\mathbf{T}x\|$, the equations imply

$$(\mathbf{T}^*\mathbf{T} - \mathbf{TT}^*)x \cdot x = 0$$

so $\mathbf{TT}^* = \mathbf{T}^*\mathbf{T}$ by Lemma C.4. $\qquad\square$

Corollary C.12. *If* $\mathbf{T} \in L(H)$ *is normal, then* $\|\mathbf{T}^2\| = \|\mathbf{T}\|^2$.

Proof. Replace x by $\mathbf{T}x$ to obtain $\|\mathbf{T}^*\mathbf{T}x\| = \|\mathbf{T}^2x\|$ so $\|\mathbf{T}^*\mathbf{T}\| = \|\mathbf{T}^2\|$. By the first part of Theorem C.11, $\|\mathbf{T}^*\mathbf{T}\| = \|\mathbf{T}\|^2$ for any $\mathbf{T} \in L(H)$. $\quad\square$

Theorem C.13. *If* $\mathbf{T} \in L(H)$ *is normal, then* $\overline{\mathcal{R}\mathbf{T}}$ *and* $\ker \mathbf{T}$ *are orthogonal complements so* $H = \overline{\mathcal{R}\mathbf{T}} \oplus \ker \mathbf{T}$.

Proof. By Theorem C.11, $\ker \mathbf{T} = \ker \mathbf{T}^*$ so the result follows from Proposition C.9. $\qquad\square$

We next establish several results which describe the spectrum of a self adjoint operator.

Theorem C.14. *Let* $\mathbf{T} \in L(H)$ *be normal. Then* $\lambda \in \rho(\mathbf{T})$ *iff there exists* $c > 0$ *such that*

$$\|(\lambda - \mathbf{T})x\| \geq c\,\|x\|$$

for all $x \in H$.

Proof. $\Longrightarrow$: It follows from Theorem B.1.

$\Longleftarrow$: If the condition is satisfied $\lambda - \mathbf{T}$ has a continuous inverse by Theorem B.1 so we must show that the range of $\lambda - \mathbf{T}$ is dense. Since $\mathbf{T}$ is normal, $\lambda - \mathbf{T}$ is normal so by Theorem C.13,

$$H = \overline{\mathcal{R}(\lambda - \mathbf{T})} \oplus \ker(\lambda - \mathbf{T}) = \overline{\mathcal{R}(\lambda - \mathbf{T})} \oplus \{0\}$$

and $\mathcal{R}(\lambda - \mathbf{T})$ is dense. $\qquad\square$

Corollary C.15. *Let* $\mathbf{T} \in L(H)$ *be normal. Then* $\lambda \in \mathbb{C}$ *is in* $\sigma(\mathbf{T})$ *iff for every* $\epsilon > 0$ *there exists* x, $\|x\| = 1$, *such that* $\|(\lambda - \mathbf{T})x\| < \epsilon$.

Recall that a point in the spectrum of a linear operator which satisfies the condition in Corollary C.15 is called an approximate eigenvalue. The set of all approximate eigenvalues is called the approximate spectrum of the operator. From Corollary C.15 it follows that the spectrum of a normal operator coincides with its approximate spectrum.

Proposition C.16. *Let* $\mathbf{T} \in L(H)$ *be normal. Then the residual spectrum of* $\mathbf{T}$, $R\sigma(\mathbf{T}) = \emptyset$.

Proof. If $\lambda \in \sigma(\mathbf{T})$ is such that $\mathcal{R}(\lambda - \mathbf{T})$ is not dense, then either $\lambda \in P\sigma(\mathbf{T})$ or $\lambda \in R\sigma(\mathbf{T})$. We show that if $\mathcal{R}(\lambda - \mathbf{T})$ is not dense, then $\lambda \in P\sigma(\mathbf{T})$. Now $\overline{\mathcal{R}(\lambda - \mathbf{T})} \neq H$ implies $\mathcal{R}(\lambda - \mathbf{T})^{\perp} \neq \{0\}$ and Theorem C.13 implies

$$\mathcal{R}(\lambda - \mathbf{T})^{\perp} = \ker(\lambda - \mathbf{T}) \neq \{0\}$$

so $\lambda \in P\sigma(\mathbf{T})$. $\qquad\square$

Recall the spectral radius of a continuous linear operator $\mathbf{T}$ is defined to be

$$r(\mathbf{T}) = \sup\{|\lambda| : \lambda \in \sigma(\mathbf{T})\}.$$

From Corollary C.12 we can obtain a formula for the spectral radius of a normal operator.

Theorem C.17. *If* $\mathbf{T} \in L(H)$ *is normal, then*

$$r(\mathbf{T}) = \|\mathbf{T}\|.$$

Proof. By Theorem B.19, $r(\mathbf{T}) = \lim \sqrt[n]{\|\mathbf{T}^n\|}$. Since powers of normal operators are normal, we have $\|\mathbf{T}^n\| = \|\mathbf{T}\|^n$ when n is even and the result follows. $\square$

Theorem C.18. *If* $\mathbf{T} \in L(H)$ *is self adjoint, then* $\sigma(\mathbf{T}) \subset \mathbb{R}$.

Proof. Let $\lambda = a + bi$ with $b \neq 0$. Let $x \in H$ and set $y = (\lambda - \mathbf{T})x$. Then

$$y \cdot x = \lambda x \cdot x - \mathbf{T}x \cdot x$$

and

$$x \cdot y = \overline{y \cdot x} = \overline{\lambda} x \cdot x - \mathbf{T}x \cdot x$$

since $\mathbf{T}x \cdot x$ is real. Hence,

$$x \cdot y - y \cdot x = (\overline{\lambda} - \lambda)x \cdot x = -2ibx \cdot x$$

and

$$2 |b| \|x\|^2 = |x \cdot y - y \cdot x| \leq 2 \|x\| \|y\|$$

by the Cauchy-Schwarz Inequality. Thus,

$$\|y\| = \|(\lambda - \mathbf{T})x\| \geq |b| \|x\|$$

so $\lambda \in \rho(\mathbf{T})$ by Theorem C.14. $\square$

If $\mathbf{T}$ is self adjoint, the bounds of $\mathbf{T}$ are defined to be

$$m(\mathbf{T}) = \inf\{\mathbf{T}x \cdot x : \|x\| = 1\}, \quad M(\mathbf{T}) = \sup\{\mathbf{T}x \cdot x : \|x\| = 1\}.$$

Theorem C.19. *If* $\mathbf{T} \in L(H)$ *is self adjoint, then*

$$\sigma(\mathbf{T}) \subset [m(\mathbf{T}), M(\mathbf{T})].$$

Proof. Suppose $\lambda \in \mathbb{R}$ is such that $\lambda > M(\mathbf{T})$. Let $\epsilon = \lambda - M(\mathbf{T})$. Then

$$(\lambda - \mathbf{T})x \cdot x = \lambda x \cdot x - \mathbf{T}x \cdot x \geq \lambda x \cdot x - M(\mathbf{T})x \cdot x = \epsilon \|x\|^2.$$

Since

$$(\lambda - \mathbf{T})x \cdot x \leq \|(\lambda - \mathbf{T})x\| \|x\|, \|(\lambda - \mathbf{T})x\| \geq \epsilon \|x\|$$

and $\lambda \in \rho(\mathbf{T})$ by Theorem C.14.

Similarly, if $\lambda < m(\mathbf{T})$, then $\lambda \in \rho(\mathbf{T})$. $\square$

Lemma C.20. *(Generalized Schwarz Inequality) If* $\mathbf{T} \in L(H)$ *is positive, then*

$$|\mathbf{T}x \cdot y|^2 \leq (\mathbf{T}x \cdot x)(\mathbf{T}y \cdot y)$$

for $x, y \in H$.

Proof.   Since $\mathbf{T}$ is positive, the function $\{x, y\} = \mathbf{T}x \cdot y$ from $H \times H \to \mathbb{C}$ has all of the properties of an inner product except possibly $\{x, x\} = 0$ iff $x = 0$. This property is not required in many of the proofs of the Cauchy-Schwarz Inequality so the inequality above follows directly from this proof (see, for example, [Sw2] Appendix, Theorem 2). $\square$

Note that the case when $\mathbf{T} = I$ is just the usual Cauchy-Schwarz Inequality.

Theorem C.21. *If* $\mathbf{T} \in L(H)$ *is self adjoint, then both* $m(\mathbf{T})$ *and* $M(\mathbf{T})$ *belong to* $\sigma(\mathbf{T})$.

Proof.   Consider the case for $m = m(\mathbf{T})$. For $x \in H$,

$$(\mathbf{T} - m)x \cdot x \geq 0$$

so $\mathbf{T} - m \geq 0$, i.e., $\mathbf{T} - m$ is positive. Apply Lemma C.20 to $\mathbf{T} - m, x$ and $y = (\mathbf{T} - m)x$ to obtain

$$|(\mathbf{T} - m)x \cdot (\mathbf{T} - m)x|^2 \leq ((\mathbf{T} - m)x \cdot x)((\mathbf{T} - m)^2 x \cdot (\mathbf{T} - m)x)$$

so

$$\|(\mathbf{T} - m)x\|^4 \leq ((\mathbf{T} - m)x \cdot x)\,\|\mathbf{T} - m\|^3\,\|x\|^2$$

and, therefore,

$$\inf\{\|(\mathbf{T} - m)x\| : \|x\| = 1\} = 0$$

since

$$\inf\{(\mathbf{T} - m)x \cdot x) : \|x\| = 1\} = 0$$

by definition. Hence, $m \in \sigma(\mathbf{T})$ by Corollary C.15. $\square$

From Theorems C.17, C.19 and C.21, we can now obtain a formula for the spectral radius in terms of the bounds for a self adjoint operator.

Theorem C.22. *If* $\mathbf{T} \in L(H)$ *is self adjoint, then*

$$\|\mathbf{T}\| = r(\mathbf{T}) = \max\{|m(\mathbf{T})|, |M(\mathbf{T})|\} = \sup\{|\mathbf{T}x \cdot x| : \|x\| = 1\}.$$

Finally, we establish a sequential completeness property for self adjoint operators which is very analogous to a completeness property for the real numbers.

Theorem C.23. *Let* $\{\mathbf{T}_j\}$ *be a sequence of self adjoint operators satisfying* $\mathbf{T}_1 \leq \mathbf{T}_2 \leq \ldots$ *and there exists a self adjoint operator* B *such that* $\mathbf{T}_j \leq B$ *for all* j. *Then there exists a self adjoint operator* $\mathbf{T} \in L(H)$ *such that* $\mathbf{T}_j x \to \mathbf{T}x$ *for every* $x \in H$ *and* $\mathbf{T} \leq B$.

Proof. We may assume

$$0 \leq \mathbf{T}_1 \leq \mathbf{T}_2 \leq \ldots \leq \mathbf{B}.$$

For $n > m$ set $\mathbf{T}_{nm} = \mathbf{T}_n - \mathbf{T}_m \geq 0$. By the generalized Schwarz Inequality, for $x \in H$,

$$(\mathbf{T}_{nm}x \cdot \mathbf{T}_{nm}x)^2 = \|\mathbf{T}_{nm}x\|^4 \leq (\mathbf{T}_{nm}x \cdot x)(\mathbf{T}_{nm}^2 x \cdot \mathbf{T}_{nm}x).$$

Since $\mathbf{T}_{nm} \leq \mathbf{B}$, Theorem C.22 implies that $\|\mathbf{T}_{nm}\| \leq \|\mathbf{B}\|$ so

$$\|\mathbf{T}_{nm}x\|^4 = \|(\mathbf{T}_n - \mathbf{T}_m)x\|^4 \leq ((\mathbf{T}_n x \cdot x) - (\mathbf{T}_m x \cdot x)) \|\mathbf{B}\|^3 \|x^2\| . \quad \text{(C.2)}$$

The sequence $\{\mathbf{T}_n x \cdot x\}$ is real, increasing and bounded and, therefore, convergent so by (C.2), the sequence $\{\mathbf{T}_n x\}$ is Cauchy and, therefore, converges to some $\mathbf{T}x \in H$. By the Banach-Steinhaus Theorem, $\mathbf{T} \in L(H)$ and

$$\mathbf{T}x \cdot y = \lim \mathbf{T}_n x \cdot y = \lim x \cdot \mathbf{T}_n y = x \cdot \mathbf{T}y$$

implies that $\mathbf{T}$ is self adjoint. Clearly, $\mathbf{T} \leq \mathbf{B}$. $\qquad\square$

Theorem C.23 can be used to show that any positive operator has a unique square root ([DM], [BN]), but we will give a proof of this fact in Theorem 2.6 by using the spectral theorem for self adjoint operators.

Definition C.24. $\mathbf{T} \in L(H)$ is unitary iff $\mathbf{T}$ is an isometry from H onto H.

In general, a linear isometry is not unitary; consider the right shift operator on l^2 defined by $\mathbf{R} : l^2 \to l^2$,

$$\mathbf{R}\{t_1, t_2, \ldots\} = \{0, t_1, t_2, \ldots\}.$$

If H is finite dimensional linear isometries are unitary.

For example, the Fourier-Plancherel transformation on $L^2(\mathbb{R}^n)$ is a unitary operator ([HS] 21.53), and if $h \in \mathbb{R}^n$, the map

$$\mathbf{T}_h : L^2(\mathbb{R}^n) \to L^2(\mathbb{R}^n)$$

defined by

$$\mathbf{T}_h f(x) = f(x + h)$$

is unitary.

Proposition C.25. *Let* $\mathbf{T} \in L(H)$. *The following are equivalent:*

(i) $\mathbf{T}^*\mathbf{T} = I$,

(ii) $\mathbf{T}x \cdot \mathbf{T}y = x \cdot y$ for $x, y \in H$,

(iii) $\|\mathbf{T}x\| = \|x\|$ for $x \in H$ (i.e., $\mathbf{T}$ is a linear isometry).

Proof. (i) $\Longrightarrow$ (ii): $\mathbf{T}^*\mathbf{T}x \cdot y = x \cdot y = \mathbf{T}x \cdot \mathbf{T}y$.

(ii) $\Longrightarrow$ (iii): Set $x = y$ in (ii).

(iii) $\Longrightarrow$ (i): $\|\mathbf{T}x\| = \|x\|$ implies

$$\|\mathbf{T}x\|^2 = \mathbf{T}x \cdot \mathbf{T}x = \|x\|^2 = x \cdot x = \mathbf{T}^*\mathbf{T}x \cdot x$$

for all x so

$$(\mathbf{T}^*\mathbf{T} - I)x \cdot x = 0$$

for all x. Hence,

$$\mathbf{T}^*\mathbf{T} - I = 0$$

by Lemma C.4. □

Theorem C.26. $\mathbf{T} \in L(H)$ *is unitary iff*

$$\mathbf{T}\mathbf{T}^* = \mathbf{T}^*\mathbf{T} = I,$$

i.e., $\mathbf{T}^* = \mathbf{T}^{-1}$.

Proof. Assume $\mathbf{T}$ is unitary. Then $\mathbf{T}^{-1} \in L(H)$. By Proposition C.25, $\mathbf{T}^*\mathbf{T} = I$ which implies

$$\mathbf{T}^*\mathbf{T}\mathbf{T}^{-1} = \mathbf{T}^{-1} = \mathbf{T}^*$$

which implies

$$\mathbf{T}^*\mathbf{T} = \mathbf{T}\mathbf{T}^* = I.$$

If $\mathbf{T}^*\mathbf{T} = \mathbf{T}\mathbf{T}^* = I$, then $\mathbf{T}^* = \mathbf{T}^{-1}$ with $\mathbf{T}$ onto. By Proposition C.25 $\mathbf{T}$ is unitary. □

Corollary C.27. *If* $\mathbf{T} \in L(H)$ *is unitary, then* $\mathbf{T}$ *is normal and the residual spectrum* $R\sigma(\mathbf{T}) = \emptyset$.

Theorem C.28. *Let* $\mathbf{T} \in L(H)$ *be unitary. Then* $\sigma(\mathbf{T}) \subset \{z : |z| = 1\}$.

Proof. Since $\mathbf{T}$ has norm 1, $\sigma(\mathbf{T}) \subset \{z : |z| \leq 1\}$ and since $\mathbf{T}^{-1} \in L(H)$, $0 \notin \sigma(\mathbf{T})$. Suppose $z \neq 0$, $|z| < 1$. Then $1/z \in \rho(\mathbf{T}^*)$ since $\|\mathbf{T}^*\| = 1$. Now

$$z - \mathbf{T} = z\mathbf{T}(\mathbf{T}^* - 1/z)$$

so

$$(z - \mathbf{T})^{-1} = (\mathbf{T}^* - 1/z)^{-1}(z\mathbf{T})^{-1} \in L(H)$$

and $z \in \rho(\mathbf{T})$. Hence, $\sigma(\mathbf{T}) \subset \{z : |z| = 1\}$. □

We will establish the converse of Theorem C.28 for normal operators using the functional calculus for normal operators in Theorem 4.13.

Appendix D

Sesquilinear Functionals

Let X be a complex inner product space.

Definition D.1. A function $b : X \times X \to \mathbb{C}$ is a sesquilinear functional if $b(\cdot, y)$ is linear for every $y \in X$ and $\overline{b(x, \cdot)}$ is linear for every $x \in X$, i.e., b is linear in the first variable and conjugate linear in the second variable.

For example, the inner product on an inner product space is a sesquilinear functional. More generally, if $\mathbf{B} : X \to X$ is linear, then

$$b(x, y) = \mathbf{B}x \cdot y$$

defines a sesquilinear functional.

Definition D.2. A sesquilinear functional b is bounded if there exists $c \geq 0$ such that

$$\|b(x, y)\| \leq c \, \|x\| \, \|y\|$$

for every $x, y \in X$. The norm of b is defined to be

$$\|b\| = \sup\{|b(x, y)| : \|x\| \leq 1, \|y\| \leq 1\}.$$

Proposition D.3. *Let* $\mathbf{B} : X \to X$ *be linear and set*

$$b(x, y) = \mathbf{B}x \cdot y$$

for $x, y \in X$. *Then* $\mathbf{B}$ *is bounded iff* b *is bounded. In this case,* $\|b\| = \|\mathbf{B}\|$.

Proof. $\Longrightarrow$: By the Cauchy-Schwarz Inequality,

$$|b(x, y)| \leq \|\mathbf{B}x\| \, \|y\| \leq \|\mathbf{B}\| \, \|x\| \, \|y\|$$

so b is bounded and $\|b\| \leq \|\mathbf{B}\|$.

$\Longleftarrow$: If $x \in X$,

$$\|\mathbf{B}x\|^2 = \mathbf{B}x \cdot \mathbf{B}x = b(x, \mathbf{B}x) \le \|b\|\,\|x\|\,\|\mathbf{B}x\|$$

so

$$\|\mathbf{B}x\| \le \|b\|\,\|x\|$$

which implies $\|\mathbf{B}\| \le \|b\|$. $\square$

We now show that every bounded sesquilinear functional on a Hilbert space has the form of the functional in Proposition D.3.

Theorem D.4. *If b is a bounded sesquilinear functional on the Hilbert space H, there exists a unique $\mathbf{B} \in L(H)$ such that $b(x, y) = \mathbf{B}x \cdot y$ for $x, y \in H$.*

Proof.   Fix $x \in H$. Then $\overline{b(x, \cdot)}$ is a bounded linear functional on H so by the Riesz Representation Theorem, there is a unique $\mathbf{B}x \in H$ such that

$$\overline{b(x, y)} = y \cdot \mathbf{B}x$$

for all $y \in H$. Since

$$\mathbf{B}x \cdot y = b(x, y),$$

the map $\mathbf{B}$ which send x into $\mathbf{B}x$ is linear and is bounded by Proposition D.3. Uniqueness is clear. $\square$

Definition D.5. A sesquilinear functional b on X is symmetric if

$$b(x, y) = \overline{b(y, x)}$$

for all $x, y \in X$.

Proposition D.6. *Let H be a Hilbert space, $\mathbf{B} \in L(H)$ and*

$$b(x, y) = \mathbf{B}x \cdot y$$

for $x, y \in H$. Then $\mathbf{B}$ is self adjoint iff b is symmetric.

Proof. $\overline{b(y, x)} = \overline{\mathbf{B}y \cdot x} = x \cdot \mathbf{B}y$ for $x, y \in H$. $\square$

Appendix E

Tempered Distributions and the Fourier Transform

We shall establish in this appendix the basic properties of the Fourier transform of tempered distributions. These results are used in Chapter 8 to treat the Weyl calculus.

In the first section we will give some basic facts about distributions. The second section is devoted to introduce the space of rapidly decreasing $\mathcal{C}^\infty$-functions and the space of tempered distributions. In the next section the Fourier transform is developed. The final sections deal with extensions of the Paley-Wiener theorem to rapidly decreasing $\mathcal{C}^\infty$-functions and the tempered distributions with compact support. We will work all the time with $\mathbb{R}^n$ although everything can be done in $\Omega \subset \mathbb{R}^n$ where Ω is an open set.

E.1 Distributions

We will start with some notation. A multi index is $\alpha = (\alpha_1, \ldots, \alpha_n) \in \mathbb{N}^n$; we define $|\alpha| = \alpha_1 + \alpha_2 + \cdots + \alpha_n$ and $\alpha! = \alpha_1!\alpha_2!\ldots\alpha_n!$. For $x \in \mathbb{R}^n$ or $\mathbb{C}^n$, $x^\alpha = x_1^{\alpha_1}x_2^{\alpha_2}\cdots x_n^{\alpha_n}$. If a and x are in $\mathbb{R}^n$ or $\mathbb{C}^n$ we define $a \cdot x = a_1x_1 + a_2x_2 + \cdots + a_nx_n$.

$\mathbf{D}^\alpha$ denotes $\mathbf{D}^\alpha = \frac{\partial^{|\alpha|}}{\partial x_1^{\alpha_1}\partial x_2^{\alpha_2}\ldots\partial x_n^{\alpha_n}}$. Finally $\alpha \leq \beta$ if and only if $\alpha_i \leq \beta_i$ for $i = 1, 2, \ldots, n$.

We will frequently use the next lemma.

Lemma E.1. *For $k \in \mathbb{N}$ there exists a constant $c \geq 0$ depending on k and n such that for $\lambda \in \mathbb{C}^n$ we have:*

$$c(1 + |\lambda|^2)^k \leq \sup_{|\beta| \leq k} |\lambda^\beta|^2 \leq \sup_{|\beta| \leq 2k} |\lambda^\beta| \leq (1 + |\lambda|^2)^{2k}$$

where $|\lambda|$ denotes the modulus of λ.

179

Proof.   For $|\beta| \leq 2k$,

$$|\lambda^{2\beta}| = |\lambda_1|^{2\beta_1} \cdots |\lambda_n|^{2\beta_n} \leq |\lambda|^{2\beta_1} \cdots |\lambda|^{\ 2\beta_n} = |\lambda|^{2|\beta|} \leq |\lambda|^{4k} \leq (1+|\lambda|^2)^{2k}.$$

On the other hand using the multinomial Newton's formula we have

$$(1 + |\lambda|^2)^k = (1 + |\lambda_1|^2 + \cdots + |\lambda_n|^2)^k$$

$$= \sum_{|\alpha| \leq k} \frac{k!}{\alpha!(k - |\alpha|)!} |\lambda_1|^{2\alpha_1} \cdots |\lambda_n|^{2\alpha_n}$$

$$= \sum_{|\alpha| \leq k} \frac{k!}{\alpha!(k - |\alpha|)!} |\lambda^{2\alpha}|$$

$$\leq \left(\sum_{|\alpha| \leq k} \frac{k!}{\alpha!(k - |\alpha|)!} \right) \sup_{\alpha \leq k} |\lambda^{2\alpha}| .$$

$$\square$$

Let f be a function in $\mathcal{C}(\mathbb{R}^n)$. The support of f is define by

$$supp(f) = cl\{ x \in \mathbb{R}^n : f(x) \neq 0 \} , \qquad (E.1)$$

that is the closure of the set of points where f is nonzero. $\mathscr{D}(\mathbb{R}^n)$ denotes the space of all infinitely differentiable functions defined on $\mathbb{R}^n$ and having a compact support. If $\mathcal{U}$ is an open set in $\mathbb{R}^n$ we define $\mathscr{D}(\mathcal{U})$ as the space of all infinitely differentiable functions defined on $\mathcal{U}$ and having a compact support contained in $\mathcal{U}$.

The space

$$\mathscr{D}_K(\mathbb{R}^n) = \{ f \in \mathcal{C}^\infty(\mathbb{R}^n) : supp(f) \subset K,$$
$$\text{where } K \subset \mathbb{R}^n, \text{ is a fixed compact set} \} . \qquad (E.2)$$

We have $\mathscr{D}(\mathbb{R}^n) = \bigcup_{K \subset \mathbb{R}^n} \mathscr{D}_K(\mathbb{R}^n)$ where K is a compact set in $\mathbb{R}^n$. If we define θ in the following way:

$$\theta(x) = \begin{cases} 0 & \text{if } \|x\| \geq 1 \\ \gamma \exp\left(\frac{1}{\|x\|^2 - 1} \right) & \text{if } \|x\| < 1 , \end{cases} \qquad (E.3)$$

where γ is a nonzero constant. The function θ is an element of $\mathscr{D}(\mathbb{R}^n)$. It is clear that the support of θ is the compact unit ball in $\mathbb{R}^n$. θ is evidently $\mathcal{C}^\infty$ in the complement of the unit sphere. Now to see that it is $\mathcal{C}^\infty$ in the whole space $\mathbb{R}^n$, take $x_0 \in \mathbb{R}^n$ such that $\|x_0\| = 1$. If x tends to x_0, with x in the interior of the unit ball, then each partial derivative of f tends to 0

since it is the product of the exponential by a rational functional function that appears in the subsequent derivations.

We can choose γ in such a way that $\int_{\mathbb{R}^n} \theta(x)dx = 1$.

Each space $\mathscr{D}_K(\mathbb{R}^n)$ has the topology induced by the family

$$p_{K,m}(f) = \sup_{x \in K} \sup_{|\beta| \leq m} \left| \mathbf{D}^\beta f(x) \right| .$$

We equip $\mathscr{D}(\mathbb{R}^n)$ with the strongest locally convex topology for which all the canonical injections $i \colon \mathscr{D}_K(\mathbb{R}^n) \longrightarrow \mathscr{D}(\mathbb{R}^n)$ are continuous. That is V is an open set in $\mathscr{D}(\mathbb{R}^n)$ if $i^{-1}(V)$ is an open set in $\mathscr{D}_K(\mathbb{R}^n)$. The topology in $\mathscr{D}(\mathbb{R}^n)$ is an inductive limit topology ([Sw2], Proposition 1, p.267) or ([Yo], p.28). A linear form on $\mathscr{D}(\mathbb{R}^n)$ will be continuous if its restriction to each $\mathscr{D}_K(\mathbb{R}^n)$ is continuous ([Sw2] p.268). The convergence of a sequence $(f_j) \subset \mathscr{D}(\mathbb{R}^n)$ to zero means that, the following two conditions are satisfied:

(i) There exists a compact subset K of $\mathbb{R}^n$ such that $supp(f_j) \subset K$ for $j \in \mathbb{N}$.

(ii) For any differential operator $\mathbf{D}^\alpha$, the sequence $(\mathbf{D}^\alpha f_j(x))$ converges uniformly to 0 on K.

Definition E.2. A linear functional T on $\mathscr{D}(\mathbb{R}^n)$ which is continuous (with respect to the topology described) is called a distribution, and is denoted $T \in \mathscr{D}'(\mathbb{R}^n)$.

Example E.3. An example of a distribution is the Dirac measure . Each $a \in \mathbb{R}^n$ determines a linear functional on $\mathscr{D}(\mathbb{R}^n)$, denoted by δ_a and defined by $\delta_a(\varphi) = \varphi(a)$ for $\varphi \in \mathscr{D}(\mathbb{R}^n)$. δ_a is called the Dirac measure on $\mathbb{R}^n$. We remark that δ_a is continuous since if K is a compact set in $\mathbb{R}^n$ containing a, we have

$$|\delta_a(\varphi)| = |\varphi(a)| \leq \max_{x \in K} |\varphi(x)| \qquad \forall \varphi \in \mathscr{D}(K)$$

and by [Sw2], Proposition 7, p.46 or [Yo], Proposition 1, p.42 we have the result.

Example E.4. Let $f \in C(\mathbb{R}^n)$ and $\varphi \in \mathscr{D}(\mathbb{R}^n)$. Since the function $f\varphi$ is continuous and vanishes outside the support of φ, then $f\varphi$ has compact support. It follows that $\int_{\mathbb{R}^n} f\varphi$ exists and $T_f(\varphi) = \int_{\mathbb{R}^n} f\varphi$ is a linear form in $\mathscr{D}(\mathbb{R}^n)$ and it is also continuous since if $\varphi \in \mathscr{D}_K(\mathbb{R}^n)$, then

$$|T_f(\varphi)| \leq \int_{\mathbb{R}^n} |f\varphi| \leq c \max_{x \in K} |f(x)| \cdot \max_{x \in K} |\varphi(x)|.$$

So if $f \in C(\mathbb{R}^n)$, T_f is a distribution.

Example E.5. Let f be a locally integrable function on $\mathbb{R}^n$, that is, f is integrable over every compact subset with respect to Lebesgue measure. Define the linear functional on $\mathscr{D}(\mathbb{R}^n)$ by $T_f(\varphi) = \int_{\mathbb{R}^n} f\varphi$. For every $\varphi \in \mathscr{D}_K(\mathbb{R}^n)$ we have

$$|T_f(\varphi)| = \left| \int_K f\varphi \right| \le \int_K |f||\varphi| \le \int_K |f| \cdot \sup_{x \in K} |\varphi(x)|$$

and by [Yo], Proposition 1, p.42 we have that T_f is continuous, so T_f is a distribution.

Sometimes $T_f(\varphi)$ is denoted by $\langle T_f, \varphi \rangle$.

Example E.6. Take δ' defined on $\mathscr{D}(\mathbb{R}^n)$ by

$$\delta'(\varphi) = \delta\left(-\frac{\partial \varphi}{\partial x}\right) = -\frac{\partial \varphi}{\partial x}(0) \quad \text{for} \quad \varphi \in \mathscr{D}(\mathbb{R}^n) \, .$$

This example can be extended to the case of more variables in the following way:

$$\partial_k \delta(\varphi) = \delta\left(-\frac{\partial}{\partial x_k}\varphi\right) = -\frac{\partial \varphi}{\partial x_k}(0) \quad \text{where} \quad 1 \le k \le n \, .$$

Later we will see that $\partial_k \delta$ is precisely the partial derivative of the Dirac measure in the sense of distributions.

Example E.7. Let μ be a bounded Borel (signed) measure on $\mathbb{R}^n$. Then μ induces a distribution T_μ by $T_\mu(\varphi) = \int_{\mathbb{R}^n} \varphi \, d\mu$ for $\varphi \in \mathscr{D}(\mathbb{R}^n)$. If $\varphi \in \mathscr{D}_K$, then $T_\mu(\varphi) \le |\mu|(K) \sup\{ |\varphi(x)| : x \in K \}$, where $|\mu|$ is the variation of μ and so T_μ is a distribution.

Example E.8. The function of one variable $\dfrac{1}{x}$ is not locally integrable on $\mathbb{R}$, hence it does not define a distribution on $\mathbb{R}$ in the way it was done in Example E.5. By definition we set

$$T_{PV\frac{1}{x}}(\varphi) = PV \int_{-\infty}^{\infty} \frac{\varphi(x)}{x} \, dx = \lim_{\varepsilon \to 0} \int_{|x| \ge \varepsilon} \frac{\varphi(x)}{x} \, dx \quad \text{for} \quad \varphi \in \mathscr{D}(\mathbb{R}) \, ,$$

where the limit on the right-hand side is called the Cauchy principal value of the integral $\int_{-\infty}^{\infty} \frac{\varphi(x)}{x} \, dx$. We have

$$\int_{|x| \ge \varepsilon} \frac{\varphi(x)}{x} \, dx = \int_{-\infty}^{-\varepsilon} \frac{\varphi(x)}{x} \, dx + \int_{\varepsilon}^{\infty} \frac{\varphi(x)}{x} \, dx$$

$$= \varphi(-\varepsilon) \ln(\varepsilon) - \int_{-\infty}^{-\varepsilon} \varphi'(x) \ln|x| dx - \varphi(\varepsilon) \ln \varepsilon - \int_{\varepsilon}^{\infty} \varphi'(x) \ln(x) dx \, .$$

Writing $\varphi(x) = \varphi(0) + x\psi(x)$ with $\psi(0) = \varphi'(0)$ and replacing above we get

$$\int_{|x|\geq\varepsilon} \frac{\varphi(x)}{x} dx = -2\varepsilon(\psi(\varepsilon) + \psi(-\varepsilon))\ln\varepsilon - \int_{-\infty}^{-\varepsilon} \varphi'(x)\ln|x|dx - \int_{\varepsilon}^{\infty} \varphi'(x)\ln(x)dx \ .$$

Now taking limits we obtain

$$\lim_{\varepsilon\to 0} \int_{|x|\geq\varepsilon} \frac{\varphi(x)}{x} dx = -\int_{-\infty}^{\infty} \varphi'(x)\ln|x|dx \ ,$$

where the last integral converges and defines a continuous linear functional on $\mathscr{D}_K(\mathbb{R})$. Hence,

$$T_{PV\frac{1}{x}}(\varphi) = \lim_{\varepsilon\to 0} \int_{|x|\geq\varepsilon} \frac{\varphi(x)}{x} dx = -\int_{-\infty}^{\infty} \varphi'(x)\ln|x|dx$$

defines a distribution on $\mathbb{R}$.

Now we want to define the product of a distribution by a C^∞-function. The motivation is the following. If $\psi \in C^\infty(\mathbb{R}^n)$ and f is locally integrable on $\mathbb{R}^n$, then ψf is locally integrable and therefore induces a distribution as follows:

$$T_{\psi f}(\varphi) = \int_{\mathbb{R}^n} \psi f\varphi = T_f(\psi\varphi) \quad \text{for} \quad \varphi \in \mathscr{D}(\mathbb{R}^n).$$

We are implicitly using the fact that $\psi\varphi \in \mathscr{D}(\mathbb{R}^n)$ when $\psi \in C^\infty(\mathbb{R}^n)$ and $\varphi \in \mathscr{D}(\mathbb{R}^n)$.

Definition E.9. Let $\psi \in C^\infty(\mathbb{R}^n)$ and $T \in \mathscr{D}'(\mathbb{R}^n)$. Then ψT is a distribution defined by $\psi T(\varphi) = T(\psi\varphi)$, or $\langle \psi T, \varphi \rangle = \langle T, \psi\varphi \rangle$.

In fact in the definition we need to prove that ψT is a distribution. The linearity is immediate. To prove that ψT is continuous it is sufficient to see that the map $T \longrightarrow \psi T$ is the transpose of the map $\varphi \longrightarrow \psi\varphi$ where $\varphi \in \mathscr{D}(\Omega)$. This last map is continuous [Sw2], Proposition 17, p.356.

Perhaps the most important property of distributions is that one can define a linear map from $\mathscr{D}'(\mathbb{R}^n)$ to $\mathscr{D}'(\mathbb{R}^n)$ which generalizes the usual partial derivative. To see how we need to define that map use $f \in C^1(\mathbb{R}^n)$ and let T_f be the distribution associated with f defined in Example E.4. Now it seems reasonable to require the distribution $\partial_j T_f$, where ∂_j denotes the partial derivative with respect to x_j, to be associated with the function $\partial_j f$.

If $\varphi \in \mathscr{D}(\mathbb{R}^n)$, then integrating by parts with respect to x_j and using that φ has compact support, we have:

$$\int_{\mathbb{R}^n} \partial_j f \cdot \varphi = -\int_{\mathbb{R}^n} f\partial_j\varphi \ .$$

If the above requirement is satisfied we can rewrite this equation as

$$\partial_j T(\varphi) = -T(\partial_j \varphi).$$

Before giving a formal definition we must prove that $\partial_j T$ is a distribution. The linearity is immediate. Now T is a distribution so it is continuous and for every compact K there exist c, m such that $|T(\varphi)| \leq cp_{K,m}(\varphi)$ for $\varphi \in \mathscr{D}_K(\mathbb{R}^n)$. Then

$$|\partial_j T(\varphi)| = |-T(\partial_j \varphi)| \leq cp_{K,m}(\partial_j \varphi) \leq cp_{K,m+1}(\varphi)$$

and therefore $\partial_j T \in \mathscr{D}'(\mathbb{R}^n)$.

By induction the following definition makes sense.

Definition E.10. Let $\alpha \in \mathbb{N}^n$, we denote by $\mathbf{D}^\alpha T$ the distributional derivative of order α of T where $T \in \mathscr{D}'(\mathbb{R}^n)$. We define $\mathbf{D}^\alpha T$ by

$$\mathbf{D}^\alpha T(\varphi) = (-1)^{|\alpha|} T(\mathbf{D}^\alpha \varphi) \quad \text{for} \quad \varphi \in \mathscr{D}(\mathbb{R})$$

or

$$\langle \mathbf{D}^\alpha T, \varphi \rangle = (-1)^{|\alpha|} \langle T, \mathbf{D}^\alpha \varphi \rangle.$$

We list a few very useful properties of the derivatives of distributions:

(1) The derivative of a distribution is a operator defined everywhere in $\mathscr{D}'(\mathbb{R}^n)$ that is, $\mathbf{D}^\alpha : \mathscr{D}'(\mathbb{R}^n) \longrightarrow \mathscr{D}'(\mathbb{R}^n)$.

(2) Every distribution has derivatives of all orders.

(3) For every $T \in \mathscr{D}'(\mathbb{R}^n)$ we have $\dfrac{\partial^2 T}{\partial x_j \partial x_i} = \dfrac{\partial^2 T}{\partial x_i \partial x_j}$.

Example E.11. The function $\log|x|$ is locally integrable in $\mathbb{R}$, hence it defines a distribution, Example E.5, whose derivative is by definition:

$$\left\langle \frac{d}{dx} \log|x|, \varphi \right\rangle = -\int_{-\infty}^{\infty} \log|x| \varphi'(x)\, dx \text{ for every } \varphi \in \mathscr{D}(\mathbb{R}).$$

Hence $\dfrac{d}{dx} \log|x| = PV\left(\dfrac{1}{x}\right)$ (see Example E.8).

Example E.12. Let's consider the Heaviside function in $\mathbb{R}$:

$$H(x) = \begin{cases} 1 & \text{if } x \geq 0 \\ 0 & \text{if } x < 0. \end{cases}$$

Its derivative in the sense of distributions is defined as:

$$\left\langle \frac{dH}{dx}, \varphi \right\rangle = -\left\langle H, \frac{d\varphi}{dx} \right\rangle = -\int_0^\infty \frac{d\varphi}{dx} = \varphi(0) = \langle \delta_0, \varphi \rangle$$

for every $\varphi \in \mathscr{D}(\mathbb{R})$. Therefore, $H' = \delta_0$ the Dirac measure. We remark that the Heaviside function is not continuous at the origin but has derivatives of all order (see Example E.6) in the sense of distributions.

Let U be an open set in $\mathbb{R}^n$, every function in $\mathscr{D}(U)$ can be consider as a function in $\mathscr{D}(\mathbb{R}^n)$. If $T \in \mathscr{D}'(\mathbb{R}^n)$, then T can be restricted to $\mathscr{D}(U)$ and denoted by $T|_U$. If $T|_U(\varphi) = 0$ for every $\varphi \in \mathscr{D}(U)$ we say that T is zero in U.

Definition E.13. Let $T \in \mathscr{D}'(\mathbb{R}^n)$ and W be the largest open subset where T is zero. The complement of W in R^n is the support of T denoted *supp T*.

The following theorem shows that the definition of support for distributions makes sense.

Theorem E.14. T *is zero in* $W = \bigcup_{j \in J} U_j$, *where* T *is zero in each open set* U_j.

Proof.   Let $\{g_j\}$ be a partition of the unity g_j in W [Ho] p.168 or [Sw2], Definition 23, p.360, subordinate to U_j. For each $\varphi \in \mathscr{D}(W)$ take $\varphi_j = g_j \varphi$. Then $\varphi_j \in \mathscr{D}(U_j)$ and since $\{U_j\}_{j \in J}$ is an open cover of the support of φ which is compact, we can select a finite subcovering $U_{j_1}, \ldots, U_{j_r}$ of the support of φ. We have:

$$g_{j_i} \in \mathscr{D}(W), \quad 0 \le g_{j_i} \le 1 \quad \text{and} \quad \sum_{i=1}^{r} g_{j_i} = 1.$$

Hence,

$$\varphi = \sum_{i=1}^{r} \varphi g_{j_i}.$$

Since $\varphi \cdot g_{j_i} \in \mathscr{D}(U_{j_i})$ and T is zero in U_{j_i}, $T(\varphi) = \sum_{i=1}^{r} T(\varphi g_{j_i}) = 0$ which implies that since φ is an arbitrary element of $\mathscr{D}(W)$, T is zero in W. $\square$

Example E.15. If $f \in \mathcal{C}(\mathbb{R}^n)$ and T_f is the corresponding distribution (see Example E.4), then *supp* $T_f = $ *supp f*.

Example E.16. The Dirac measure has compact support equal to $\{0\}$.

Later we will need a characterization of the distributions with compact support. We first define a locally convex topology on $\mathcal{C}^\infty(\mathbb{R}^n)$. Let $j \in \mathbb{N}$ and $K \subset \mathbb{R}^n$ be compact. Set

$$p_{j,K}(f) = \sup\{ |\mathbf{D}^\alpha f(x)| : |\alpha| \le j, x \in K \}$$

and give $C^\infty(\mathbb{R}^n)$ the locally convex topology generated by the semi-norms, $p_{j,K}$ where $j \in \mathbb{N}$ and $K \subset \mathbb{R}^n$ is compact. Thus, a sequence $\{\varphi_\ell\} \subset C^\infty(\mathbb{R}^n)$ converges to 0 iff $\mathbf{D}^\alpha \varphi_\ell \longrightarrow 0$ uniformly on compact sets of $\mathbb{R}^n$ for every α.

Theorem E.17. *The injection $\mathscr{D}(\mathbb{R}^n) \subset C^\infty(\mathbb{R}^n)$ is continuous and $\mathscr{D}(\mathbb{R}^n)$ is dense in $C^\infty(\mathbb{R}^n)$.*

Proof. The continuity is clear. Let $\varphi \in C^\infty(\mathbb{R}^n)$. Let $\psi_j \in \mathscr{D}(\mathbb{R}^n)$ be equal to 1 in $\{x : \|x\| \leq j\}$, $\psi_j = 0$ in $\{x : \|x\| > j+1\}$ and $0 \leq \psi_j \leq 1$. Then $\psi_j \longrightarrow 1$ in $C^\infty(\mathbb{R}^n)$ and $\psi_j \varphi \longrightarrow \varphi$ in $C^\infty(\mathbb{R}^n)$ and $\psi_j \varphi \in \mathscr{D}(\mathbb{R}^n)$. $\square$

Since the injection of $\mathscr{D}(\mathbb{R}^n)$ into $C^\infty(\mathbb{R}^n)$ is continuous and has dense range, its transpose is a one-one map [Ho. ex 5, p. 320] so the dual $(C^\infty(\mathbb{R}^n))'$ can be identified as linear subspace of $\mathscr{D}'(\mathbb{R}^n)$. We show that $(C^\infty(\mathbb{R}^n))'$ can be identified with the distributions of compact support.

Theorem E.18. *A distribution $T \in \mathscr{D}'(\mathbb{R}^n)$ has compact support iff $T \in (C^\infty(\mathbb{R}^n))'$.*

Proof. The proof is based on the observation that a distribution T belongs to $(C^\infty(\mathbb{R}^n))'$ iff $\varphi \longrightarrow \langle T, \varphi \rangle$ is continuous on $\mathscr{D}(\mathbb{R}^n)$ with respect to the topology induced on $\mathscr{D}(\mathbb{R}^n)$ by the topology of $C^\infty(\mathbb{R}^n)$ since if this linear functional is continuous, it has a unique continuous linear extension on $C^\infty(\mathbb{R}^n)$.

First, suppose $T \in (C^\infty(\mathbb{R}^n))'$. Then there exists a compact $K \subset \mathbb{R}^n$, $j \geq 0$ and $c > 0$ such that $|\langle T, \varphi \rangle| \leq cp_{j,K}\varphi$ for $\varphi \in C^\infty(\mathbb{R}^n)$. Then $\langle T, \varphi \rangle = 0$ for $\varphi \in \mathscr{D}(\mathbb{R}^n)$, $supp(\varphi) \bigcap K = \emptyset$. Thus, T vanishes outside K.

Next, suppose $K = supp(T)$ is compact. Pick $\psi \in \mathscr{D}(\mathbb{R}^n)$ such that $\psi = 1$ on a neighborhood of K so $\psi T = T$. Each $\psi \varphi \in \mathscr{D}(\mathbb{R}^n)$ has support contained in $L = supp\, \psi$. On $\mathscr{D}_L(\mathbb{R}^n)$ the topologies induced by $\mathscr{D}(\mathbb{R}^n)$ and $C^\infty(\mathbb{R}^n)$ coincide so if a net $\{\varphi_l\}$ converges to 0 in $C^\infty(\mathbb{R}^n)$, then $\{\psi \varphi_l\}$ converges to 0 in $\mathscr{D}(\mathbb{R}^n)$ and $\langle T, \psi \varphi_l \rangle = \langle \psi T, \varphi_l \rangle = \langle T, \varphi_l \rangle \longrightarrow 0$. Thus, $\varphi \longrightarrow \langle T, \varphi \rangle$ is continuous on $\mathscr{D}(\mathbb{R}^n)$ with respect to the topology induced by $C^\infty(\mathbb{R}^n)$. $\square$

E.2 The Spaces $\mathcal{S}(\mathbb{R}^n)$ and $\mathcal{S}'(\mathbb{R}^n)$

Definition E.19. $\mathcal{S}(\mathbb{R}^n)$ is the space of rapidly decreasing $\mathcal{C}^\infty$-functions sometimes called the Schwartz space and defined by

$$\mathcal{S}(\mathbb{R}^n) = \{\, f \in \mathcal{C}^\infty(\mathbb{R}^n) : \sup_{x \in \mathbb{R}^n} |x^\alpha \mathbf{D}^\beta f(x)| = \|x^\alpha \mathbf{D}^\beta f\|_\infty < \infty, \forall \alpha, \beta \in \mathbb{N}^n \,\}\,.$$

A function f is in $\mathcal{S}(\mathbb{R}^n)$ if $\lim_{|x| \to \infty} |x^\alpha \mathbf{D}^\beta f(x)| = 0$ and that is why sometimes $\mathcal{S}(\mathbb{R}^n)$ is also called the space of all $\mathcal{C}^\infty$-functions rapidly decreasing at infinity.

Evidently $\mathcal{S}(\mathbb{R}^n)$ is a vector subspace of $\mathcal{C}^\infty(\mathbb{R}^n)$ and

$$q_{\alpha,\beta}(f) = \|x^\alpha \mathbf{D}^\beta f\|_\infty = \sup_{x \in \mathbb{R}^n} |x^\alpha \mathbf{D}^\beta f(x)| \quad \text{for all } \alpha, \beta \in \mathbb{N}^n$$

is a semi-norm in $\mathcal{S}(\mathbb{R}^n)$. We give $\mathcal{S}(\mathbb{R}^n)$ the locally convex topology induced by the semi-norms $q_{\alpha,\beta}$, where $\alpha, \beta \in \mathbb{N}^n$.

The next theorem characterizes the functions belonging to $\mathcal{S}(\mathbb{R}^n)$.

Theorem E.20.

(a) *For $f \in \mathcal{C}^\infty(\mathbb{R}^n)$ the following are equivalent:*

    (i) $f \in \mathcal{S}(\mathbb{R}^n)$.

    (ii) $\|x^\alpha \mathbf{D}^\beta f\|_1 = \displaystyle\int_{\mathbb{R}^n} |x^\alpha \mathbf{D}^\beta f(x)| dx < \infty$ *for every $\alpha, \beta \in \mathbb{N}^n$.*

    (iii) $\displaystyle\sup_{|\beta| \leq m} \left|(1+|x|^2)^k \mathbf{D}^\beta f\right|_\infty = \sup_{|\beta| \leq m} \sup_{x \in \mathbb{R}^n} \left|(1+|x|^2)^k \mathbf{D}^\beta f(x)\right| < \infty$
    for every $k, m \in \mathbb{N}$.

(b) *The topology in $\mathcal{S}(\mathbb{R}^n)$ induced by the family $q_{\alpha,\beta}(f) = \|x^\alpha \mathbf{D}^\beta f\|_\infty$ with $\alpha, \beta \in \mathbb{N}^n$ is equivalent to the one induced by:*

    (i) $p_{\alpha,\beta}(f) = \|x^\alpha \mathbf{D}^\beta f\|_1$ *with $\alpha, \beta \in \mathbb{N}^n$.*

    (ii) $r_{k,m}(f) = \sup_{|\beta| \leq m} \|(1 + |x|^2)^k \mathbf{D}^\beta f\|_\infty$ *for $k, m \in \mathbb{N}$.*

Proof. For $f \in \mathcal{C}^\infty(\mathbb{R}^n), k, m \in \mathbb{N}$ define:

$$s_{k,m}(f) = \sup_{|\alpha| \leq k} \sup_{|\beta| \leq m} p_{\alpha,\beta}(f).$$

The theorem will follow from the inequalities:

$$cr_{k,m}(f) \leq \sup_{\substack{|\alpha| \leq 2k \\ |\beta| \leq m}} q_{\alpha,\beta}(f) \leq r_{k,m}(f), \tag{E.4}$$

$$s_{k,m}(f) \leq \pi^k r_{k,m}(f), \tag{E.5}$$

$$\sup_{\substack{|\alpha| \leq 2k \\ |\beta| \leq m}} q_{\alpha,\beta}(f) \leq s_{k,m+k}. \tag{E.6}$$

Inequality (E.4) is immediate from Lemma E.1.

To prove inequality (E.5) we write $f(x) = (1 + |x|^2)^k f(x) \frac{1}{(1+|x|^2)^k}$. Since

$$\int_{\mathbb{R}^n} \frac{dx}{(1 + |x|^2)^k} \leq \pi^n, \text{ then } \left| \int_{\mathbb{R}^n} f(x) dx \right| \leq \pi^n r_{k,0}(f).$$

From Lemma E.1 $\sup_{|\alpha| \leq k} |x^\alpha| \leq (1 + |x|^2)^k$ so $\sup_{|\alpha| \leq k} \|x^\alpha f\| \leq r_{k,0}(f)$ and then we can see that inequality (E.5) holds.

To prove inequality (E.6) it is enough to do it for $k = m = 0$. Let $f \in \mathcal{C}^\infty(\mathbb{R}^n)$ such that $s_{k,0}(f) < \infty$. The equality:

$$f(x) - f(a) = \int_{a_1}^{x_1} \cdots \int_{a_n}^{x_n} \frac{\partial^n}{\partial x_1 \cdots \partial x_n} f(x) \, dx$$

holds since $f \in \mathcal{C}^\infty(\mathbb{R}^n)$.

Since $s_{k,0}(f) < \infty$,

$$\lim_{a_i \to -\infty} \int_{a_1}^{x_1} \cdots \int_{a_n}^{x_n} \mathbf{D}^n f(x) \, dx$$

exists and using the previous equality we have that $\lim_{a_i \to -\infty} f(a)$ also exists.

Again using $s_{k,0}(f) < \infty$ we can conclude that $\lim_{a_i \to -\infty} f(a)$ is zero. Suppose not, then there exists $M \neq 0$ such that $\lim_{a_i \to -\infty} |f(a)| = M$. This means that for every $\varepsilon > 0$ there exists N such that if $a_i < N$ then $\|f(a)| - M\| < \varepsilon$.

Take $\varepsilon = \frac{M}{2}$ and we will have $\frac{M}{2} < |f(a)|$ then

$$\int_{a_1}^{x_1} \cdots \int_{-\infty}^{N} \cdots \int_{a_n}^{x_n} \frac{M}{2} \leq \int_{a_1}^{x_1} \cdots \int_{-\infty}^{N} \cdots \int_{a_n}^{x_n} |f(x)| \, dx \ .$$

The first integral is not bounded and so the second will not be bounded so this contradicts the fact that $s_{k,0}(f) < \infty$ and then we have $\lim_{a_i \to -\infty} f(a) = 0$.

Then

$$f(x) = \int_{a_1}^{x_1} \cdots \int_{a_n}^{x_n} \frac{\partial^n}{\partial x_1 \cdots \partial x_n} f(x) dx,$$

we can conclude

$$\|f\|_\infty \leq \left\| \frac{\partial^n}{\partial x_1 \cdots \partial x_n} f \right\|_1 \leq s_{k,0}(f).$$

$\square$

Example E.21. The space $\mathcal{D}(\mathbb{R}^n)$ is a vector subspace of $\mathcal{S}(\mathbb{R}^n)$.

Example E.22. The function $\exp\left(-\frac{|x|^2}{2}\right)$ belongs to $\mathcal{S}(\mathbb{R}^n)$.

Example E.23. Let $\varphi \in \mathscr{D}(\mathbb{R}^n)$ be such that $0 \leq \varphi \leq 1$, *supp* φ is a subset of the unit ball and $\varphi(0) = 1$. Let (x_j) be a sequence of elements of $\mathbb{R}^n$ such that $|x_j| + 2 \leq |x_{j+1}|$ and define

$$\psi(x) = \sum_{j=1}^{\infty} \frac{\varphi(x - x_j)}{(1 + |x|^2)^j}.$$

The sum is well define since the support of the functions $\varphi(x - x_j)$ are disjoint. If $\alpha \in \mathbb{N}$ and $\beta \in \mathbb{N}^n$, we have

$$(1 + |x|^2)^{\alpha} \mathbf{D}^{\beta} \psi(x) = \frac{(1 + |x|^2)^{\alpha} \mathbf{D}^{\beta} \varphi(x - x_j)}{(1 + |x|^2)^{\alpha} (1 + |x|^2)^{j - \alpha}}$$

whenever $|x_j| - 1 \leq |x| \leq |x_j| + 1$. On the other hand $\frac{1 + |x|^2}{1 + |x_j|^2} \leq c$ for a suitable constant and $\sup_{x \in \mathbb{R}^n} |\mathbf{D}^{\beta} \varphi(x - x_j)| = \sup_{x \in \mathbb{R}^n} |\mathbf{D}^{\beta} \varphi(x)|$. It follows that $\psi \in \mathcal{S}(\mathbb{R}^n)$.

Remark E.24.

(a) $\mathcal{S}(\mathbb{R}^n)$ is a subspace of $\mathcal{C}_0(\mathbb{R}^n) = \{f \in \mathcal{C}(\mathbb{R}^n): \text{ for every } \varepsilon > 0, \text{ there exists } K_{f,\varepsilon} \text{ a compact subset of } \mathbb{R}^n \text{ such that } |f(x)| < \varepsilon \text{ for } x \in \mathbb{R}^n \backslash K_{f,\varepsilon}\}$ and the canonical injection is continuous since for $f \in \mathcal{S}(\mathbb{R}^n)$, $\|f\|_{\infty} \leq q_{0,0}(f)$, where the norm in $\mathcal{C}_0 \mathbb{R}^n)$ is $\| \cdot \|_{\infty}$.
(b) Clearly $\mathcal{S}(\mathbb{R}^n) \subset L^1(\mathbb{R}^n)$ with a continuous injection.
(c) $\mathcal{S}(\mathbb{R}^n) \subset \mathcal{C}^{\infty}(\mathbb{R}^n)$ and the canonical injection is also continuous where the topology in $\mathcal{C}^{\infty}(\mathbb{R}^n)$ is given by $p_{K,m}(f) = \sup\limits_{|\alpha| \leq m} \sup\limits_{x \in K} |\mathbf{D}^{\alpha} f(x)|$.
(d) $\mathscr{D}_K(\mathbb{R}^n) \subset \mathcal{S}(\mathbb{R}^n)$; in $\mathscr{D}_K(\mathbb{R}^n)$ we have the topology given by $p_{K,m}(f) = \sup\limits_{|\alpha| \leq m} \sup\limits_{x \in K} |\mathbf{D}^{\alpha} f(x)|$. Now $\mathscr{D}(\mathbb{R}^n) = \bigcup \mathscr{D}_K(\mathbb{R}^n) \subset \mathcal{S}(\mathbb{R}^n)$ and since for each K the injection from $\mathscr{D}_K(\mathbb{R}^n)$ to $\mathcal{S}(\mathbb{R}^n)$ is continuous, then the injection from $\mathscr{D}(\mathbb{R}^n)$ to $\mathcal{S}(\mathbb{R}^n)$ is also continuous.
(e) $\mathbf{D}^{\alpha} f \in \mathcal{S}(\mathbb{R}^n)$ when $f \in \mathcal{S}(\mathbb{R}^n)$ and the map $f \longrightarrow \mathbf{D}^{\alpha} f$ is continuous.

Theorem E.25. $\mathscr{D}(\mathbb{R}^n)$ *is a dense subspace of* $\mathcal{S}(\mathbb{R}^n)$.

Proof. Choose $f \in \mathcal{S}(\mathbb{R}^n)$. Now take $\psi \in \mathscr{D}(\mathbb{R}^n)$ so that ψ equals 1 on the closed unit ball of $\mathbb{R}^n$ and for every $j \in \mathbb{N}$ define $\psi_j(x) = \psi\left(\frac{x}{j}\right)$. The sequence (ψ_j) has the following properties:

(i) $\psi_j \in \mathscr{D}(\mathbb{R}^n)$ for every $j \in \mathbb{N}$.
(ii) $\psi_j - 1$ is zero in the ball with radius j and center 0.
(iii) $\sup\limits_{j \in \mathbb{N}} \sup\limits_{x \in \mathbb{R}^n} |\mathbf{D}^{\beta} \psi_j(x)| \leq \sup\limits_{x \in \mathbb{R}^n} |\mathbf{D}^{\beta} \psi(x)|$ for all $\beta \in \mathbb{N}^n$.

Put $f_j(x) = f(x)\psi\left(\frac{x}{j}\right)$ for $x \in \mathbb{R}^n$ and $j \in \mathbb{N}$. From (i) $f_j \in \mathscr{D}\mathbb{R}^n)$. We want to prove that $f_j \longrightarrow f$ in $\mathcal{S}(\mathbb{R}^n)$ with respect to the topology induced by $q_{\alpha,\beta}(f) = \left\| x^\alpha \mathbf{D}^\beta f \right\|_\infty$. Let $g_j = f - f_j = f(1 - \psi_j)$, we need to prove that $q_{\alpha,\beta}(g_j) \to 0$.

$$\mathbf{D}^\beta g_j = \sum_{\gamma \leq \beta} \frac{\beta!}{(\beta - \gamma)!\gamma!} \mathbf{D}^{\beta-\gamma}(1 - \psi_j)\mathbf{D}^\gamma f.$$

By (ii) $\mathbf{D}^{\beta-\gamma}(1 - \psi_j)$ is zero for every $\beta \in \mathbb{N}^n$ and $|x| \leq j$ and since $f \in \mathcal{S}(\mathbb{R}^n)$ we have that $x^\alpha \mathbf{D}^\gamma f$ is bounded. It follows that $q_{\alpha,\beta}(g_j) \to 0$ uniformly on $\mathbb{R}^n$ when $j \longrightarrow \infty$. Thus $f_j \longrightarrow f$ in $\mathcal{S}(\mathbb{R}^n)$ and the density is proved. $\qquad\square$

We have then that $i: \mathscr{D}(\mathbb{R}^n) \longrightarrow \mathcal{S}(\mathbb{R}^n)$ is continuous and that $\mathscr{D}(\mathbb{R}^n)$ is dense in $\mathcal{S}(\mathbb{R}^n)$. If F is a continuous linear functional on $\mathcal{S}(\mathbb{R}^n)$ we define $T = F \circ i$. Then corresponding to each F we have a T. Furthermore different F's cannot define the same T since $\mathscr{D}(\mathbb{R}^n)$ is dense in $\mathcal{S}(\mathbb{R}^n)$. So we have an isomorphism between $\mathcal{S}'(\mathbb{R}^n)$ the dual of $\mathcal{S}(\mathbb{R}^n)$ and a subset of $\mathscr{D}'(\mathbb{R}^n)$. This subset is the space of tempered distributions on $\mathbb{R}^n$ and is formed by those T in $\mathscr{D}'(\mathbb{R}^n)$ that can be extended in a unique form to be a continuous functional defined on $\mathcal{S}(\mathbb{R}^n)$. In general we will identify F with T and we will say that the *tempered distributions* are the elements of $\mathcal{S}'(\mathbb{R}^n)$.

Definition E.26. The elements of $\mathcal{S}'(\mathbb{R}^n)$ are said to be tempered distributions.

Example E.27. A distribution with compact support is a tempered distribution, since the topology induced in $\mathscr{D}(\mathbb{R}^n)$ by $\mathcal{S}(\mathbb{R}^n)$ is finer than the topology induced by $\mathcal{C}^\infty(\mathbb{R}^n)$. We then have $(\mathcal{C}^\infty(\mathbb{R}^n))' \subset \mathcal{S}'(\mathbb{R}^n)$. (Remark E.24 (c) and (d).)

Example E.28. Every function $f \in L^p(\mathbb{R}^n), 1 \leq p \leq \infty$, defines a tempered distribution by setting

$$\langle f, \varphi \rangle = \int_{\mathbb{R}^n} f\varphi \, dx \quad \text{for every } \varphi \in \mathcal{S}(\mathbb{R}^n).$$

Indeed we have $|\langle f, \varphi \rangle| \leq \|f\|_p\|\varphi\|_q$ with $\frac{1}{p} + \frac{1}{q} = 1$ by Holder's inequality. By observing that if a sequence (φ_j) converges to zero in $\mathcal{S}(\mathbb{R}^n)$ it also converges to zero in L^q, we conclude that every $f \in L^p$ defines a continuous linear functional in $\mathcal{S}(\mathbb{R}^n)$. Moreover, it can be seen that $L^p(\mathbb{R}^n)$ can be identified with a vector subspace of $\mathcal{S}'(\mathbb{R}^n)$.

Example E.29. Every polynomial $p(x)$ with constant coefficients defines a tempered distribution. In fact it suffices to show that every monomial x^α defines a tempered distribution by setting

$$\langle x^\alpha, \varphi \rangle = \int_{\mathbb{R}^n} x^\alpha \varphi(x)\, dx \quad \text{for every } \varphi \in \mathcal{S}(\mathbb{R}^n).$$

Example E.30. A continuous function $f(x)$ is slowly increasing at infinity if there exists an integer $k \geq 0$ such that $(1+x^2)^{-\frac{k}{2}} f(x)$ is bounded in $\mathbb{R}^n$. Or equivalently, if there is a polynomial $p(x)$ such that $|f(x)| \leq |p(x)|$ for every $x \in \mathbb{R}^n$.

Every continuous function f slowly increasing at infinity defines a tempered distribution. Set:

$$\langle f, \varphi \rangle = \int_{\mathbb{R}^n} f(x)\varphi(x)\, dx \quad \text{for every } \varphi \in \mathcal{S}(\mathbb{R}^n).$$

We have:

$$
\begin{aligned}
|\langle f, \varphi \rangle| &\leq \int_{\mathbb{R}^n} |f(x)\varphi(x)|\, dx \\
&= \int_{\mathbb{R}^n} |(1+x^2)^{-\frac{k}{2}} f(x)(1+x^2)^{\frac{k}{2}} \varphi(x)|\, dx \\
&\leq c \int_{\mathbb{R}^n} |(1+x^2)^{\frac{k}{2}} \varphi(x)|\, dx.
\end{aligned}
$$

By observing that if a sequence (φ_j) converges to zero in $\mathcal{S}(\mathbb{R}^n)$, then for every $k \geq 0, ((1+x^2)^{\frac{k}{2}} \varphi_j)$ converges to zero in L^1 (Theorem E.20 (bii)). We conclude that f defines an element of $\mathcal{S}'(\mathbb{R}^n)$.

E.3 Fourier Transform of Functions

In what follows $x \cdot \xi = \xi \cdot x = x_1 \xi_1 + \cdots + x_n \xi_n$, for $x, \xi \in \mathbb{R}^n$.

Definition E.31. Let $f \in L^1(\mathbb{R}^n)$. The Fourier transform of f denoted by $\mathscr{F}(f)$ or $\hat{f}$ is defined by

$$(\mathscr{F}f)(\xi) = \int_{\mathbb{R}^n} f(x) e^{-2\pi i x \xi}\, dx, \quad \xi \in \mathbb{R}^n .$$

(a) It is clear that the Fourier transform is linear.

(b) If $g(x) = f(kx)$ for $k \in \mathbb{R}$ then a change of variable in the definition of $\hat{f}$ gives $(\mathscr{F}g)(\xi) = \frac{1}{|k|^n}(\mathscr{F}f)(\frac{\xi}{k})$.

(c) Let $f(x_1, \ldots, x_n) = f_1(x_1) \cdot f_2(x_2) \cdots f_n(x_n)$ where $(x_1, \ldots, x_n) \in \mathbb{R}^n$ and each $f_i \in L^1(\mathbb{R})$. Then $\mathscr{F}(f) = \mathscr{F}(f_1) \cdot \mathscr{F}(f_2) \cdots \mathscr{F}(f_n)$ can be proved by directly applying the definition and the Fubini-Tonelli Theorem.

Example E.32. Let $f : \mathbb{R}^n \longrightarrow \mathbb{R}$ be such that $f(x) = e^{-\pi |x|^2}$. Then $\mathscr{F}(f) = f$.

Proof. It is clear that $f \in \mathcal{S}(\mathbb{R}^n)$. It is sufficient to examine the case $n = 1$ by (c). We have:

$$\widehat{f}(\xi) = \int_{\mathbb{R}^n} e^{-\pi x^2 - 2\pi i \xi x} \, dx = \int_{\mathbb{R}^n} e^{-\pi (x + i\xi)^2 - \pi \xi^2} \, dx$$
$$= e^{-\pi \xi^2} \int_{\mathbb{R}^n} e^{-\pi (x + i\xi)^2} \, dx \ .$$

Consider the function $z \longrightarrow \exp(-\pi z^2)$. This is an entire function of the variable $z = x + iy$. So by the integral Cauchy formula its integral along the rectangle with vertices $-a, a, a + i\eta, -a + i\eta$ with $a > 0$ is zero. The integral in the vertical sides tends to 0 if $a \longrightarrow \infty$ since:

$$\left| \int_a^{a + i\eta} e^{(-\pi z^2)} \, dz \right| = \left| i \int_0^\eta e^{-\pi (a^2 - y^2)} e^{-2\pi i a y} \, dy \right|$$
$$\leq |\eta| e^{-\pi (a^2 - \eta^2)}.$$

Now we have:

$$\int_{-\infty}^\infty e^{-(x + i\xi)^2} \, dx = \lim_{a \to \infty} \int_{-a + i\eta}^{a + i\eta} e^{-\pi z^2} \, dz = \lim_{a \to \infty} \int_{-a}^a e^{-\pi z^2} \, dz$$
$$= \int_{-\infty}^\infty e^{-\pi x^2} \, dx = 1 \ .$$

It follows that $\widehat{f} = f$ and $\widehat{f}(0) = \int_{\mathbb{R}} f(x) \, dx$, then $f(0) = \int_{\mathbb{R}} \widehat{f}(x) \, dx$. $\qquad \square$

The next theorem accounts for the fundamental properties of Fourier transform.

Theorem E.33.

(a) *Let $f \in L^1(\mathbb{R}^n)$. Then:*

    (i) *$\widehat{f}$ is a continuous and bounded function in $\mathbb{R}^n$. Furthermore, $\|\widehat{f}\|_\infty \leq \|f\|_1$.*

    (ii) *$\widehat{f}(\xi)$ tends to 0 if $|\xi| \longrightarrow \infty$ and then $\widehat{f} \in C_0(\mathbb{R}^n)$ (Riemann–Lebesgue lemma).*

(b) If $f, g \in L^1(\mathbb{R}^n)$, then $\int_{\mathbb{R}^n} \widehat{f}(\xi) g(\xi) \, d\xi = \int_{\mathbb{R}^n} f(x) \widehat{g}(x) \, dx.$

(c)   *(i)* If $x^\alpha f \in L^1(\mathbb{R}^n)$ *for every* $\alpha \in \mathbb{N}^n$ *and* $|\alpha| \leq k$, *then* $\widehat{f} \in C^k(\mathbb{R}^n)$ *and* $\mathbf{D}^\alpha(\widehat{f}) = \widehat{(-2\pi i x)^\alpha f}.$

     (ii) If $f \in C^k(\mathbb{R}^n)$ *and* $\mathbf{D}^\beta(f) \in L^1(\mathbb{R})$ *for every* $\beta \in \mathbb{N}^n$ *and* $|\beta| \leq k$, *then* $x^\beta \widehat{f} \in L^\infty(\mathbb{R}^n)$, *furthermore* $(2\pi i x)^\beta \widehat{f} = \widehat{\mathbf{D}^\beta f}.$

Proof.   (ai) Let $f \in L^1(\mathbb{R}^n)$. The function $\xi \longrightarrow f(x) e^{-2\pi i \xi x}$ is continuous for every $\xi \in \mathbb{R}^n$, and is bounded by $|f(x)|$. Using the Lebesgue dominated convergence theorem and the continuity of the integrand, $\widehat{f}(\xi)$ is continuous in $\mathbb{R}^n$, and $|\widehat{f}(\xi)| \leq \int_{\mathbb{R}^n} |f(x)| \, dx$, which proves that $\widehat{f}$ is bounded in $\mathbb{R}^n$ and the property of the norms.

The proof of (aii) will be done after the proof of (cii).

(b) Let $f, g \in L^1(\mathbb{R}^n)$, by (ai) $\widehat{f}$ and $\widehat{g} \in L^\infty(\mathbb{R})$. Using Holder's inequality $\widehat{f} g$ and $f \widehat{g}$ are in $L^1(\mathbb{R}^n)$. f, g are integrable functions on $\mathbb{R}^n$ so $f(x) \cdot g(y)$ is integrable on $\mathbb{R}^n \times \mathbb{R}^n$ and the function $(x, y) \longrightarrow f(x) g(y) e^{-2\pi i x y}$ is integrable on $\mathbb{R}^n \times \mathbb{R}^n$.

By Fubini's theorem:

$$\int_{\mathbb{R}^n \times \mathbb{R}^n} f(x) g(y) e^{-2\pi i x y} \, dx = \int_{\mathbb{R}^n} f(x) \left(\int_{\mathbb{R}^n} g(y) e^{-2\pi i x y} dy \right) dx$$

$$= \int_{\mathbb{R}^n} g(y) \left(\int_{\mathbb{R}^n} f(x) e^{-2\pi i x y} dx \right) dy \, .$$

The second part of this equality is $\int f(x) \widehat{g}(x) \, dx$ and the third part is $\int g(y) \widehat{f}(y) \, dy.$

(ci) It will be enough to prove that if the function $x_1 \longrightarrow x_1 f(x)$ is integrable, then $\widehat{f}$ is differentiable (with respect to ξ_1) and $\dfrac{\partial}{\partial \xi_1} \widehat{f}(\xi) = \int_{\mathbb{R}^n} -2\pi i x_1 f(x) e^{-2\pi i x \xi} \, dx.$

Observe that the function $\xi \longrightarrow -2\pi i x_1 e^{-2\pi i x \xi} f(x)$ is continuous on $\mathbb{R}^n$ and it is dominated by $2\pi |x_1 f(x)|$ so we can differentiate under the integration sign and we will get the result.

(cii) We will prove that if $\dfrac{\partial f}{\partial x_1} \in L^1(\mathbb{R}^n) \cap C(\mathbb{R}^n)$, then:

$$2\pi i \xi_1 \widehat{f}(\xi) = \int_{\mathbb{R}^n} \frac{\partial f(x)}{\partial x_1} e^{-2\pi i x \xi} dx \, .$$

To do this:

$$\int_{\mathbb{R}^n} \left(\frac{\partial f}{\partial x_1}(x)e^{-2\pi ix\xi} - 2\pi i\xi_1 f(x)e^{-2\pi ix\xi} \right) dx$$

$$= \int_{\mathbb{R}^n} \frac{\partial}{\partial x_1} \left(f(x)e^{-2\pi ix\xi} \right) dx$$

$$= \int_{\mathbb{R}^{n-1}} f(x)e^{-2\pi ix\xi} \Big|_{-\infty}^{\infty} dx_2 \cdots dx_n \ .$$

Now we will prove that $f(-\infty) = f(+\infty) = 0$. Since $\dfrac{\partial f}{\partial x_1} \in L^1(\mathbb{R}^n)$, then

$$\lim_{a\to\infty} \int_0^a \frac{\partial f}{\partial x_1} dx_1 \text{ and } \lim_{a\to-\infty} \int_a^0 \frac{\partial f}{\partial x_1} dx_1 \text{ both exist and since } \frac{\partial f}{\partial x_1} \in C(\mathbb{R}^n),$$

then $f(a) - f(0) = \displaystyle\int_0^a \frac{\partial f}{\partial x_1} dx_1$. Then $f(+\infty)$ and $f(-\infty)$ exist and since $f \in L^1(\mathbb{R}^n)$, a similar argument to the one given in Theorem E.20 and inequality (E.6) proves that $f(+\infty)$ and $f(-\infty)$ are zero.

(aii) We need to prove the Riemann–Lebesgue lemma. Suppose that $f \in \mathscr{D}(\mathbb{R}^n)$. By (cii) we have:

$$\widehat{\frac{\partial}{\partial x_i} f}(\xi) = 2i\pi\xi_i \widehat{f}(\xi) \qquad i = 1, \ldots, n \ .$$

For $\xi_i \neq 0$ we have:

$$|\widehat{f}(\xi)| = \frac{1}{2\pi|\xi_i|} \left| \widehat{\frac{\partial}{\partial x_i} f}(\xi) \right| \le \frac{1}{2\pi|\xi_i|} \left\| \frac{\partial}{\partial x_i} f \right\|_\infty \ .$$

By (ai) we have $\left\| \widehat{\dfrac{\partial}{\partial x_i} f} \right\|_\infty \le \left\| \dfrac{\partial}{\partial x_i} f \right\|_1$ so when $|\xi| \to \infty$ at least one of the $|\xi_i| \to \infty$ and then $\widehat{f}(\xi) \to 0$.

The general case is done using the density of $\mathscr{D}(\mathbb{R}^n)$ in $L^1(\mathbb{R}^n)$. Let $f \in L^1(\mathbb{R}^n)$, there exists (f_j) a sequence in $\mathscr{D}(\mathbb{R}^n)$ such that $f_j \to f$ in the norm of $L^1(\mathbb{R}^n)$. Since $\left\| \widehat{f} - \widehat{f_j} \right\|_\infty \le \|f_j - f\|_1$, the sequence $(\widehat{f_j})$ converges uniformly in $\mathbb{R}^n$ to $\widehat{f}$, but every $\widehat{f_j}$ belongs to $C_0(\mathbb{R}^n)$ and since $C_0(\mathbb{R}^n)$ is closed in the uniform convergence topology we have that $\widehat{f} \in C_0(\mathbb{R}^n)$. $\square$

Corollary E.34. *If $f, g \in L^1(\mathbb{R}^n)$ then*

$$\int_{\mathbb{R}^n} \widehat{f}(\xi)g(\xi)e^{2i\pi a\xi} d\xi = \int_{\mathbb{R}^n} f(x+a)\widehat{g}(x) dx \ .$$

Proof. By Theorem E.33 (b) applied to $f(x+a)$ and $\widehat{g}(x)$ we have:

$$\int_{\mathbb{R}^n} f(x+a)\widehat{g}(x)\,dx = \int_{\mathbb{R}^n} \widehat{f}(x+a)g(x)\,dx$$

$$= \int_{\mathbb{R}^n} \left(\int_{\mathbb{R}^n} f(t+a)e^{-2\pi i\xi\cdot t}\,dt\right) g(\xi)\,d\xi$$

$$= \int_{\mathbb{R}^n} \left(\int_{\mathbb{R}^n} f(x)e^{-2\pi i\xi\cdot(x-a)}\,dx\right) g(\xi)\,d\xi$$

$$= \int_{\mathbb{R}^n} \widehat{f}(\xi)e^{2\pi i\xi a}g(\xi)\,d\xi .$$

$\square$

We arrive now to one of the main theorems of this theory.

Theorem E.35. *The Fourier transform $\mathscr{F}$ is an isomorphism from $\mathcal{S}(\mathbb{R}^n)$ to $\mathcal{S}(\mathbb{R}^n)$. Let $\overline{\mathscr{F}}$ be defined by:*

$$\overline{\mathscr{F}}(f)(\xi) = \int_{\mathbb{R}^n} f(x)e^{2\pi i\xi x}\,dx \quad \text{for every } \xi \in \mathbb{R}^n .$$

Then $\overline{\mathscr{F}}$ is the inverse isomorphism that is:

$$\mathscr{F}\overline{\mathscr{F}}(f) = \overline{\mathscr{F}}\mathscr{F}(f) = f \quad \text{for every } f \in \mathcal{S}(\mathbb{R}^n) .$$

Proof.

Claim E.36. *$\mathscr{F}$ is linear and continuous from $\mathcal{S}(\mathbb{R}^n)$ to $\mathcal{S}(\mathbb{R}^n)$.*

The linearity is immediate.
For $\alpha, \beta \in \mathbb{N}^n$, Theorem E.33 (c) indicates:

$$(2i\pi x)^\alpha \mathbf{D}^\beta \widehat{f} = (2i\pi x)^\alpha \overline{(-2\pi ix)^\beta f} = \mathbf{D}^\alpha \overline{[(-2\pi ix)^\beta f]} .$$

From Theorem E.33 (a) we have:

$$\|x^\alpha \mathbf{D}^\beta \widehat{f}\|_\infty \le (2\pi)^{|\beta|-|\alpha|} \|\mathbf{D}^\alpha x^\beta f\|_1 .$$

This last inequality proves that if $f \in \mathcal{S}(\mathbb{R}^n)$, then $\widehat{f} \in \mathcal{S}(\mathbb{R}^n)$ and that the Fourier transform is continuous; in fact the map $f \longrightarrow \mathbf{D}^\alpha(x^\beta f)$ is continuous since it is the composition of two continuous maps and $f \longrightarrow \|f\|_1$ is also continuous on $\mathcal{S}(\mathbb{R}^n)$ with the topology given by $p_{0,0}$ defined in Theorem E.20 (bi).

Claim E.37. *$\overline{\mathscr{F}}$ is linear and continuous from $\mathcal{S}(\mathbb{R}^n)$ to $\mathcal{S}(\mathbb{R}^n)$.*

$\mathscr{F}$ and $\overline{\mathscr{F}}$ differ only by i and $-i$ so Claim E.36 also proves that $\overline{\mathscr{F}}$ is continuous.

Claim E.38. $\overline{\mathscr{F}}$ *is the inverse of* $\mathscr{F}$.

We will only prove that $\overline{\mathscr{F}}\mathscr{F}(f) = f$ that is:

$$\int_{\mathbb{R}^n} \left(\int_{\mathbb{R}^n} f(x)e^{-2\pi i\xi x}\, dx \right) e^{2\pi i\xi a}\, d\xi = f(a)\ .$$

Now since the function $(x, \xi) \longrightarrow f(x)e^{2\pi i\xi(a-x)}$ need not be integrable on $\mathbb{R}^n \times \mathbb{R}^n$, Fubini's theorem cannot be applied.

Let $\varphi \in \mathcal{S}(\mathbb{R}^n)$ such that $\varphi(0) = 1$ and define $g_j(x) = \varphi(\frac{x}{j})$ for $j \in \mathbb{N}$. By Corollary E.34 and Remark E.24(b) we have:

$$\int_{\mathbb{R}^n} \widehat{f}(\xi)g_j(\xi)e^{2\pi i a\xi}\, d\xi = \int_{\mathbb{R}^n} f(a+x)\widehat{g_j(x)}\, dx = \int_{\mathbb{R}^n} f(a + \tfrac{s}{j})\widehat{\varphi}(s)\, ds\ .$$

If $j \to \infty$, $\widehat{f}(\xi)g_j(\xi)e^{2\pi i a\xi} \to \widehat{f}(\xi)\varphi(0)e^{2\pi i a\xi}$; we have $g_j(x) = \varphi(\frac{x}{j})$ so it is easy to prove that:

$$|\widehat{f}(\xi)g_j(\xi)e^{2\pi i a\xi}| \le \|\varphi\|_\infty |\widehat{f}(\xi)|\ .$$

By the Lebesgue dominated convergence theorem:

$$\int_{\mathbb{R}^n} \widehat{f}(\xi)g_j(\xi)e^{2\pi i a\xi}\, d\xi \longrightarrow \int_{\mathbb{R}^n} \widehat{f}(\xi)\varphi(0)e^{2\pi i a\xi}\, d\xi\ .$$

Furthermore we have:

$$\left| f\left(a + \frac{s}{j}\right)\widehat{\varphi}(s) \right| \le \|f\|_\infty |\widehat{\varphi}(s)|$$

and

$$f\left(a + \frac{s}{j}\right)\widehat{\varphi}(s) \longrightarrow f(a)\widehat{\varphi}(s)\ .$$

So again using the Lebesgue dominated convergence theorem:

$$\int_{\mathbb{R}^n} f(a + \tfrac{s}{j})\widehat{\varphi}(s)\, ds \longrightarrow \int_{\mathbb{R}^n} f(a)\widehat{\varphi}(s)\, ds\ .$$

Then

$$\int_{\mathbb{R}^n} \varphi(0)\widehat{f}(\xi)e^{2\pi i a\xi}\, d\xi = \int_{\mathbb{R}^n} f(a)\widehat{\varphi}(s)\, ds \qquad (\text{E.7})$$

and this is true for any $\varphi \in \mathcal{S}(\mathbb{R}^n)$ such that $\varphi(0) = 1$ so pick $\varphi(x) = e^{-\pi|x|^2}$. By Example E.22: $\widehat{\varphi}(\xi) = e^{-2\pi|\xi|^2}$ and since $\varphi(0) = 1$ we have that $\int_{\mathbb{R}^n} \widehat{\varphi}(s)\, ds = 1$ and then by (E.7):

$$\varphi(0)\int_{\mathbb{R}^n} \widehat{f}(\xi)e^{2\pi i a\xi}\, d\xi = f(a)\int_{\mathbb{R}^n} \widehat{\varphi}(s)\, ds$$

and

$$f(a) = \int_{\mathbb{R}^n} \widehat{f}(\xi)e^{2\pi i a\xi}\, d\xi\ . \qquad\qquad \square$$

Theorem E.39. *If $f, g \in \mathcal{S}(\mathbb{R}^n)$ we have that $\int_{\mathbb{R}^n} \widehat{f} \cdot g = \int_{\mathbb{R}^n} f \cdot \widehat{g}.$*

Proof. From Fubini's theorem we have:

$$
\int_{\mathbb{R}^n} \widehat{f}(\xi) g(\xi) e^{2\pi i x \xi} \, d\xi = \int_{\mathbb{R}^n} g(\xi) e^{2\pi i x \xi} \left(\int_{\mathbb{R}^n} e^{-2\pi i y \xi} f(y) \, dy \right) d\xi
$$

$$
= \int_{\mathbb{R}^n} \int_{\mathbb{R}^n} g(\xi) f(y) e^{2\pi i (x-y)\xi} \, d\xi \, dy
$$

$$
= \int_{\mathbb{R}^n} \widehat{g}(y - x) f(y) \, dy
$$

for all $f, g \in \mathcal{S}(\mathbb{R}^n)$.
Changing variables we can write:

$$
\int_{\mathbb{R}^n} \widehat{f}(\xi) g(\xi) e^{2\pi i x \xi} \, d\xi = \int_{\mathbb{R}^n} \widehat{g}(y) f(x + y) \, dy \, .
$$

Taking $x = 0$ we get:

$$
\int_{\mathbb{R}^n} \widehat{f}(\xi) g(\xi) \, d\xi = \int_{\mathbb{R}^n} \widehat{g}(y) f(y) \, dy \, .
$$

$\square$

E.4 Fourier Transform of a Tempered Distribution

To motivate the next definition take $f \in L^1(\mathbb{R}^n)$, $\widehat{f} \in L^\infty(\mathbb{R}^n)$. They both induce an element of $\mathcal{S}'(\mathbb{R}^n)$ and furthermore by Theorem E.39 we have

$$
\widehat{f}(\varphi) = \int \widehat{f} \varphi = \int f \widehat{\varphi} = f(\widehat{\varphi})
$$

so for $T \in \mathcal{S}'(\mathbb{R}^n)$ we can define $\mathscr{F}T$ by

$$
\langle \mathscr{F}T, \varphi \rangle = \langle T, \widehat{\varphi} \rangle
$$

and since the map $\varphi \to \widehat{\varphi}$ from $\mathcal{S}(\mathbb{R}^n)$ to $\mathcal{S}'(\mathbb{R}^n)$ is continuous, then $\mathscr{F}T \in \mathcal{S}'(\mathbb{R}^n)$. Observe that $\mathscr{F}T$ is the transpose of $\mathscr{F} : \mathcal{S}(\mathbb{R}^n) \longrightarrow \mathcal{S}(\mathbb{R}^n)$.

We have seen that the Fourier transform is an isomorphism of $\mathcal{S}(\mathbb{R}^n)$ onto $\mathcal{S}(\mathbb{R}^n)$. This allows us to define, by duality, the Fourier transform of tempered distributions.

Definition E.40. Let $T \in \mathcal{S}(\mathbb{R}^n)$. Its Fourier transform is the image of T under the transpose map: ${}^t\mathscr{F} : \mathcal{S}'(\mathbb{R}^n) \longrightarrow \mathcal{S}'(\mathbb{R}^n)$ of the map $\mathscr{F} : \mathcal{S}(\mathbb{R}^n) \longrightarrow \mathcal{S}(\mathbb{R}^n)$.

We will make here an abuse of notation and will denote ${}^t\mathscr{F}$ also by $\mathscr{F}$. Then we have:

$$\langle \mathscr{F}T, \varphi \rangle = \langle T, \mathscr{F}\varphi \rangle \text{ for } T \in S'(\mathbb{R}^n) \text{ and } \varphi \in S(\mathbb{R}^n) \ .$$

Indeed the reason we use the same $\mathscr{F}$ to represent either the Fourier transform of elements of $S(\mathbb{R}^n)$ or the Fourier transform of tempered distributions is that when $T = \varphi \in S(\mathbb{R}^n)$ then by Theorem E.39 it follows that the definition for tempered distributions coincides with the definition for functions in $S(\mathbb{R}^n)$.

Furthermore $\mathscr{F}T$ is again a tempered distribution since $\varphi \to \mathscr{F}(\varphi)$ is a continuous linear map from $S(\mathbb{R}^n)$ to $S(\mathbb{R}^n)$ and T is continuous and linear on $S(\mathbb{R}^n)$ so $\langle T, \mathscr{F}\varphi \rangle$ is continuous and linear on $S(\mathbb{R}^n)$ and then $\mathscr{F}(T) \in S'(\mathbb{R}^n)$. So for each $T \in S'(\mathbb{R}^n)$ we have that $\mathscr{F}(T) \in S'(\mathbb{R}^n)$.

Example E.41. $\mathscr{F}(\delta) = 1$ since for $\varphi \in S(\mathbb{R}^n)$ we have:

$$\langle \mathscr{F}(\delta), \varphi \rangle = \langle \delta, \mathscr{F}\varphi \rangle = \widehat{\varphi}(0) = \int_{\mathbb{R}^n} \varphi(\xi) \, d\xi = \langle 1, \varphi \rangle \ .$$

Example E.42. For $T \in S'(\mathbb{R}^n)$ the translate of T by a vector $h \in \mathbb{R}^n$ is defined by $\langle \tau_h T, \varphi \rangle = \langle T, \tau_{-h} \varphi \rangle$ for all $\varphi \in S(\mathbb{R}^n)$, where $(\tau_{-h}\varphi)(x) = \varphi(x + h)$. Since

$$\sup_{x \in \mathbb{R}^n} \left| \mathbf{D}^\beta \varphi(x) \right| = \sup_{x \in \mathbb{R}^n} \left| \mathbf{D}^\beta (\tau_{-h}\varphi)(x) \right| \quad \text{for all } \beta \in \mathbb{N}^n \ ,$$

the map $\tau_{-h} \colon S(\mathbb{R}^n) \longrightarrow S(\mathbb{R}^n)$ is an isomorphism and its transpose $\tau_h \colon S'(\mathbb{R}^n) \longrightarrow S'(\mathbb{R}^n)$ is also an isomorphism ([Sw2], Theorem 14, p.316). We have:

$$\mathscr{F}(\tau_h T) = e^{-2\pi i h} \mathscr{F}(T)$$

and

$$\mathscr{F}\left(e^{2\pi i h x} T\right) = \tau_h \mathscr{F}(T).$$

Indeed for

$$\varphi \in S(\mathbb{R}^n)$$

we have:

$$\begin{aligned}
\mathscr{F}(\tau_h \varphi)(x) &= \int_{\mathbb{R}^n} \varphi(\xi - h) e^{-2\pi i x \xi} \, d\xi \\
&= \int_{\mathbb{R}^n} \varphi(\xi) e^{-2\pi i x (\xi + h)} \, d\xi \\
&= e^{-2\pi i x h} (\mathscr{F}\varphi)(x)
\end{aligned}$$

and

$$(\tau_{-h}\mathscr{F}(\varphi))(x) = \int_{\mathbb{R}^n} \varphi(\xi)e^{-2\pi i(\xi+h)x} \, d\xi$$
$$= \mathscr{F}\left(e^{-2\pi ih\xi}\varphi(\xi)\right)(x).$$

Hence:

$$\langle \mathscr{F}(\tau_h T), \varphi \rangle = \langle T, \tau_{-h}\mathscr{F}(\varphi) \rangle$$
$$= \left\langle T, \mathscr{F}\left(e^{-2\pi ih\xi}\varphi\right) \right\rangle$$
$$= \left\langle e^{-2\pi ih\xi}\mathscr{F}(T), \varphi \right\rangle$$

which proves the first formula and:

$$\left\langle \mathscr{F}\left(e^{2\pi ihx} T\right), \varphi \right\rangle = \left\langle T, e^{2\pi ihx}\mathscr{F}(\varphi) \right\rangle$$
$$= \langle T, \mathscr{F}(\tau_{-h}\varphi) \rangle$$
$$= \langle \tau_h\mathscr{F}(T), \varphi \rangle$$

which proves the second formula.

Coming back to the general setting we have that $\mathscr{F}\colon \mathcal{S}(\mathbb{R}^n) \longrightarrow \mathcal{S}(\mathbb{R}^n)$ is continuous and the transpose is ${}^t\mathscr{F}\colon \mathcal{S}'(\mathbb{R}^n) \longrightarrow \mathcal{S}'(\mathbb{R}^n)$ is also continuous ([Sw2], Theorem 14, p.316) if $\mathcal{S}'(\mathbb{R}^n)$ has either the weak, the strong or the Mackey topology.

Now following the same arguments as in Definition E.40, we can define ${}^t\overline{\mathscr{F}}\colon \mathcal{S}'(\mathbb{R}^n) \longrightarrow \mathcal{S}'(\mathbb{R}^n)$ as the transpose of $\overline{\mathscr{F}}\colon \mathcal{S}(\mathbb{R}^n) \longrightarrow \mathcal{S}(\mathbb{R}^n)$ and again we will make an abuse of notation by using $\overline{\mathscr{F}}$ instead of ${}^t\overline{\mathscr{F}}$ so

$$\left\langle \overline{\mathscr{F}}T, \varphi \right\rangle = \left\langle T, \overline{\mathscr{F}}\varphi \right\rangle \text{ for } T \in \mathcal{S}'(\mathbb{R}^n) \text{ and } \varphi \in \mathcal{S}(\mathbb{R}^n).$$

It is also clear that $\overline{\mathscr{F}}$ is continuous. We summarize all of the above properties in the following theorem.

Theorem E.43. *The Fourier transform* $\mathscr{F}\colon \mathcal{S}'(\mathbb{R}^n) \longrightarrow \mathcal{S}'(\mathbb{R}^n)$ *is an isomorphism.* $\overline{\mathscr{F}}$ *is the inverse isomorphism* $\overline{\mathscr{F}}\colon \mathcal{S}'(\mathbb{R}^n) \longrightarrow \mathcal{S}'(\mathbb{R}^n)$ *and* $\mathscr{F}\overline{\mathscr{F}}(T) = \overline{\mathscr{F}}\mathscr{F}(T) = T.$

Proof. We only need to prove the last equality:

$$\left\langle \mathscr{F}\overline{\mathscr{F}}T, \varphi \right\rangle = \left\langle \overline{\mathscr{F}}T, \mathscr{F}\varphi \right\rangle = \left\langle T, \overline{\mathscr{F}}\mathscr{F}\varphi \right\rangle = \langle T, \varphi \rangle. \qquad \square$$

Proposition E.44. *If $T \in \mathcal{S}'(\mathbb{R}^n)$ and $\alpha \in \mathbb{N}^n$ then:*

$$\mathscr{F}(\mathbf{D}^\alpha T) = (2\pi i \xi)^\alpha \mathscr{F}(T).$$

Proof.   Take $\varphi \in \mathcal{S}(\mathbb{R}^n)$, we have:

$$\langle \mathscr{F}(\mathbf{D}^\alpha T), \varphi \rangle = \langle \mathbf{D}^\alpha T, \mathscr{F}\varphi \rangle = (-1)^{|\alpha|} \langle T, \mathbf{D}^\alpha(\mathscr{F}\varphi) \rangle .$$

By Theorem E.33 (ci) we have

$$\mathbf{D}^\alpha(\mathscr{F}\varphi) = \widehat{(-2\pi i \xi)^\alpha \varphi}.$$

Hence,

$$\langle \mathscr{F}(\mathbf{D}^\alpha T), \varphi \rangle = (-1)^{|\alpha|} \langle T, \mathbf{D}^\alpha(\mathscr{F}\varphi) \rangle$$
$$= (-1)^{|\alpha|} \langle \mathscr{F}T, (-2\pi i \xi)^\alpha \varphi \rangle = \langle (2\pi i \xi)^\alpha \mathscr{F}T, \varphi \rangle$$

and then

$$\mathscr{F}(\mathbf{D}^\alpha T) = (2\pi i \xi)^\alpha \mathscr{F}(T). \qquad \square$$

Proposition E.45. *If $T \in \mathcal{S}'(\mathbb{R}^n)$ and $\alpha \in \mathbb{N}^n$, then*

$$\mathscr{F}[(-2\pi i x)^\alpha T] = \mathbf{D}^\alpha(\mathscr{F}T) .$$

Proof.   Take $\varphi \in \mathcal{S}(\mathbb{R}^n)$, we have:

$$\langle \mathscr{F}[(-2\pi i x)^\alpha T], \varphi \rangle = \langle (-2\pi i x)^\alpha T, \mathscr{F}\varphi \rangle$$
$$= (-1)^{|\alpha|} \langle (2\pi i x)^\alpha T, \mathscr{F}\varphi \rangle$$
$$= (-1)^{|\alpha|} \langle T, (2\pi i x)^\alpha \mathscr{F}\varphi \rangle .$$

By Theorem E.33 (cii)

$$(2\pi i x)^\alpha \mathscr{F}\varphi = \mathscr{F}(\mathbf{D}^\alpha \varphi) .$$

Hence,

$$\langle \mathscr{F}[(-2\pi i x)^\alpha T], \varphi \rangle = (-1)^{|\alpha|} \langle T, \mathscr{F}(\mathbf{D}^\alpha \varphi) \rangle$$
$$= (-1)^{|\alpha|} \langle \mathscr{F}T, \mathbf{D}^\alpha \varphi \rangle$$
$$= \langle \mathbf{D}^\alpha(\mathscr{F}(T)), \varphi \rangle .$$

We conclude

$$\mathscr{F}[(-2\pi i x)^\alpha T] = \mathscr{F}(\mathbf{D}^\alpha T). \qquad \square$$

If f and g are defined on $\mathbb{R}^n$ their convolution $f * g$ is defined by $(f * g)(x) = \int_{\mathbb{R}^n} f(y) \cdot g(x - y)\, dy$ provided that the integral exists almost everywhere for $x \in \mathbb{R}^n$.

Proposition E.46. *The map* $\mathcal{S}(\mathbb{R}^n) \times \mathcal{S}(\mathbb{R}^n) \longrightarrow \mathcal{S}(\mathbb{R}^n)$ *such that to* (f, g) *we associate* $f * g$ *is well defined, bilinear and continuous.*

Proof. By using the Fubini–Tonelli Theorem and the Fourier inverse we get:

$$
\begin{aligned}
\widehat{f \cdot g}(\xi) &= \int_{\mathbb{R}^n} e^{-2\pi i x \xi} f(x) g(x)\, dx \\
&= \int_{\mathbb{R}^n} e^{-2\pi i x \xi} g(x) \left(\int_{\mathbb{R}^n} e^{2\pi i x \eta} \widehat{f}(\eta)\, d\eta \right) dx \\
&= \int_{\mathbb{R}^n} \int_{\mathbb{R}^n} e^{-2\pi i x (\xi - \eta)} g(x) \widehat{f}(\eta)\, dx\, d\eta \\
&= \int_{\mathbb{R}^n} \left(\int_{\mathbb{R}^n} e^{-2\pi i x (\xi - \eta)} g(x)\, dx \right) \widehat{f}(\eta)\, d\eta \\
&= \int_{\mathbb{R}^n} \widehat{g}(\xi - \eta) \widehat{f}(\eta)\, d\eta = \left(\widehat{f} * \widehat{g} \right)(\xi).
\end{aligned}
$$

Note that $f, g \in \mathcal{S}(\mathbb{R}^n)$; hence their Fourier transform are in $\mathcal{S}(\mathbb{R}^n)$, then $\widehat{f} * \widehat{g} \in \mathcal{S}(\mathbb{R}^n)$ and this, since the Fourier transform maps $\mathcal{S}(\mathbb{R}^n)$ onto $\mathcal{S}(\mathbb{R}^n)$, proves that $f * g \in \mathcal{S}(\mathbb{R}^n)$.

The bilinearity is immediate from the definition.

Now for $\alpha \leq \gamma$ and $|\gamma| \leq m$ the convolution $(x^{\gamma - \alpha} f) * x^\alpha g$ is well defined and $(x^{\gamma - \alpha} f * x^\alpha g)(x) = \int_{\mathbb{R}^n} (x - y)^{\gamma - \alpha} f(x - y) y^\alpha g(y)\, dy$. Now using the binomial Newton formula:

$$
[(x - y) + y]^\alpha = \sum_{\alpha \leq \gamma} \frac{\gamma!}{\alpha!(\gamma - \alpha)!} (x - y)^{\gamma - \alpha} y^\alpha
$$

and then

$$
x^\alpha (f * g)(x) = \sum_{\alpha \leq \gamma} \frac{\gamma!}{\alpha!(\gamma - \alpha)!} (x^{\gamma - \alpha} f * x^\alpha g)(x).
$$

Then there exists a constant c_ℓ such that $p_{\ell, \beta}(f * g) \leq c_\ell p_{\ell, \beta}(f) p_{\ell, 0}(g)$ which proves that $(f, g) \longrightarrow f * g$ is continuous. $\square$

Let us introduce some notation $\check{f}(x) = f(-x)$ and $(\tau_a f)(x) = f(x - a)$ (Example E.42). Now we have that for every $g \in \mathcal{S}(\mathbb{R}^n)$ the linear map

$f \longrightarrow \check{g} * f$ is a continuous map from $\mathcal{S}(\mathbb{R}^n)$ to $\mathcal{S}(\mathbb{R}^n)$. Denote by $T \longrightarrow g*T$ its transpose; that is

$$\langle g * T, f \rangle = \langle T, \check{g} * f \rangle.$$

Then $g*T$ is an element of $\mathcal{S}'(\mathbb{R}^n)$ and $T \longrightarrow g*T$ is linear and continuous from $\mathcal{S}'(\mathbb{R}^n)$ to $\mathcal{S}'(\mathbb{R}^n)$. We define $g*T$ to be the convolution of $g \in \mathcal{S}(\mathbb{R}^n)$ and $T \in \mathcal{S}'(\mathbb{R}^n)$, we also denote this product by $T * g$.

Proposition E.47. *Let $\varphi \in \mathcal{S}(\mathbb{R}^n)$ and $T \in \mathcal{S}'(\mathbb{R}^n)$. Then $\mathscr{F}(\varphi * T) = (\mathscr{F}\varphi)(\mathscr{F}T)$ and $\mathscr{F}(\varphi T) = \mathscr{F}\varphi * \mathscr{F}T$.*

Proof.   Let us first prove that the proposition holds for φ and $\psi \in \mathcal{S}(\mathbb{R}^n)$ that is

$$\overline{\mathscr{F}}(\varphi\psi) = \overline{\mathscr{F}}(\varphi) * \overline{\mathscr{F}}(\psi) \quad \text{and} \quad \mathscr{F}(\varphi * \psi) = (\mathscr{F}\varphi)(\mathscr{F}\psi).$$

By Corollary E.34

$$\int_{\mathbb{R}^n} (\mathscr{F}f)(\xi)g(\xi)e^{2\pi i a\xi}\,d\xi = \int_{\mathbb{R}^n} f(a+x)(\mathscr{F}g)(x)\,dx$$

or

$$\overline{\mathscr{F}}[(\mathscr{F}f)g] = f * \check{\mathscr{F}}g = f * \overline{\mathscr{F}}g.$$

Take $g = \psi$, $\mathscr{F}f = \varphi$ and then

$$\overline{\mathscr{F}}(\varphi\psi) = (\overline{\mathscr{F}}\varphi) * (\overline{\mathscr{F}}\psi). \tag{E.8}$$

Taking Fourier transform in both sides

$$\varphi\psi = \mathscr{F}[(\overline{\mathscr{F}}\varphi) * (\overline{\mathscr{F}}\psi)].$$

Denote $\varphi_1 = \overline{\mathscr{F}}\varphi$ and $\psi_1 = \overline{\mathscr{F}}\psi$ and then

$$(\mathscr{F}\varphi_1)(\mathscr{F}\psi_1) = \mathscr{F}(\varphi_1 * \psi_1). \tag{E.9}$$

Now let $\psi \in \mathcal{S}(\mathbb{R}^n)$ then:

$$\langle \mathscr{F}(\varphi * T), \psi \rangle = \langle \varphi * T, \mathscr{F}\psi \rangle = \langle T, \check{\varphi} * \mathscr{F}\psi \rangle$$
$$= \langle \overline{\mathscr{F}}\mathscr{F}T, \check{\varphi} * \mathscr{F}\psi \rangle = \langle \mathscr{F}T, \overline{\mathscr{F}}(\check{\varphi} * \mathscr{F}\psi) \rangle.$$

Using equation (E.9) we get

$$\overline{\mathscr{F}}(\check{\varphi} * \mathscr{F}\psi) = (\overline{\mathscr{F}}\check{\varphi})(\overline{\mathscr{F}}\mathscr{F}\psi) = (\mathscr{F}\check{\varphi})\psi,$$

now we finish the proof in the following way

$$\langle \mathscr{F}(\varphi * T), \psi \rangle = \langle \mathscr{F}T, \overline{\mathscr{F}}(\check{\varphi} * \mathscr{F}\psi) \rangle$$
$$= \langle \mathscr{F}T, (\mathscr{F}\varphi)\psi \rangle$$
$$= \langle \mathscr{F}\varphi\mathscr{F}T, \psi \rangle$$

for every $\psi \in \mathcal{S}(\mathbb{R}^n)$ so:

$$\mathscr{F}(\varphi * T) = (\mathscr{F}\varphi)(\mathscr{F}T). \tag{E.10}$$

It remains to prove that $\mathscr{F}(\varphi T) = (\mathscr{F}\varphi) * (\mathscr{F}T)$. Let $\varphi_1 \in \mathcal{S}(\mathbb{R}^n)$ and $T_1 \in \mathcal{S}'(\mathbb{R}^n)$ by equation (E.10), we have $\mathscr{F}(\varphi_1 * T_1) = (\mathscr{F}\varphi_1)(\mathscr{F}T_1)$. Take $T = \mathscr{F}T_1$ and $\varphi = \mathscr{F}\varphi_1$ and then $T_1 = \overline{\mathscr{F}}T$ and $\varphi_1 = \overline{\mathscr{F}}\varphi$ so we get

$$\mathscr{F}(\overline{\mathscr{F}}\varphi * \overline{\mathscr{F}}T) = \mathscr{F}(\overline{\mathscr{F}}\varphi) \cdot \mathscr{F}(\overline{\mathscr{F}}T) = \varphi T$$

so

$$\overline{\mathscr{F}}\varphi * \overline{\mathscr{F}}T = \overline{\mathscr{F}}(\varphi T)$$

which is equivalent to what was needed to prove. $\qquad\square$

Proposition E.48. *Let R be a positive number and let $T \in \mathscr{D}'(\mathbb{R}^n)$ such that $supp\, T \subset \{\, x : |x| < R\,\}$. Then $(\mathscr{F}T)(\xi) = \big\langle T(x), e^{-2\pi i \xi x} \big\rangle$ for $\xi \in \mathbb{R}^n$.*

Proof. We first observe that the right-hand side of the equality in the proposition is well defined because T is a distribution with compact support while $e^{-2\pi i \xi x}$ is a C^∞-function of x.

Take ψ in $\mathcal{S}(\mathbb{R}^n)$ such that $\psi \equiv 1$ on a neighborhood of $\{\, x : |x| \leq R\,\}$. By Proposition E.47 and the fact that $\psi T = T$ we have

$$\mathscr{F}T = \mathscr{F}(\psi T) = \mathscr{F}\psi * \mathscr{F}T.$$

Take $\varphi \in \mathcal{S}(\mathbb{R}^n)$ such that $\mathscr{F}\varphi = \psi$ then:

$$\begin{aligned}
(\mathscr{F}T)(\xi) = (\mathscr{F}T * \mathscr{F}\psi)(\xi) = (\mathscr{F}T * \check{\varphi})(\xi) &= \mathscr{F}T(\tau_x \varphi) \\
&= T(\mathscr{F}(\tau_x \varphi)) = T(e^{-2\pi i \xi x} \mathscr{F}\varphi) \\
&= T(\psi e^{-2\pi i \xi x}) = T(e^{-2\pi i \xi x})
\end{aligned}$$

and then

$$(\mathscr{F}T)(\xi) = \big\langle T(x), e^{-2\pi i \xi x} \big\rangle. \qquad\square$$

E.5 Paley–Wiener Theorems

The Fourier transform of a function in $\mathscr{D}(\mathbb{R}^n)$ is characterized by the Paley–Wiener Theorem for functions. The second Theorem in this section will be the generalization of the Paley–Wiener Theorem for distributions with compact support.

Theorem E.49 (Paley–Wiener for functions in $\mathscr{D}(\mathbb{R}^n)$). *Let R be a positive number. The following are equivalent for a function f defined on $\mathbb{R}^n$:*

(i) *f is the image of the Fourier transform of a function $\varphi \in \mathscr{D}(\mathbb{R}^n)$ with supp $\varphi \subset \{x : |x| \leq R\}$.*
(ii) *f can be extended to a function $\tilde{f}$ analytic on $\mathbb{C}^n$ such that for every $k \in \mathbb{N}$ there exists c_k such that:*

$$|\tilde{f}(\zeta)| \leq c_k \left(1 + |\zeta|^2\right)^{-k} e^{2\pi R |\mathrm{Im}(\zeta)|} \quad \text{for} \quad \zeta \in \mathbb{C}^n .$$

We remark that c_k does not depend on ζ but can depend on f, and

$$|\mathrm{Im}(\zeta)| = \left[\sum_{j=1}^{n} |\mathrm{Im}(\zeta_j)|^2 \right]^{1/2} .$$

Proof. (i)$\Rightarrow$(ii) : Let $f = \mathscr{F}\varphi$ with $\varphi \in \mathscr{D}(\mathbb{R}^n)$ and supp $\varphi \subset \{x : |x| \leq R\}$.
Define for all $\zeta \in \mathbb{C}^n$

$$\tilde{f}(\zeta) = \int_{\mathbb{R}^n} \varphi(x) e^{-2i\pi x \zeta}\, dx$$

where $x\zeta = x_1\zeta_1 + \cdots + x_n\zeta_n$.
If $x \in \{x : |x| \leq R\}$, then if $\zeta = a + ib$

$$\left| e^{-2i\pi x \zeta} \right| = \left| e^{-2i\pi x(a+ib)} \right| = e^{2\pi x b} \leq e^{2\pi R |b|} .$$

The integrand is therefore in $L^1(\mathbb{R}^n)$ for every $\zeta \in \mathbb{C}^n$ and $\tilde{f}$ is well defined on $\mathbb{C}^n$. So $\tilde{f}$ is clearly an extension of f. Since $\mathbf{D}^\beta \left(\varphi(x) e^{-2i\pi x \zeta} \right) \in L^1(\mathbb{R}^n)$ for $\beta \in \mathbb{N}^n$, then

$$\left(\mathbf{D}^\beta \tilde{f} \right)(\zeta) = \int_{\mathbb{R}^n} (-2\pi i x)^\beta \varphi(x) e^{-2i\pi x \zeta}\, dx \; \text{for} \beta \in \mathbb{N}^n.$$

This shows $\tilde{f}$ is $\mathcal{C}^\infty$ and then applying the Cauchy–Riemann equation we have that $\tilde{f}$ is entire.
Using the same type of arguments as in Theorem E.33 (cii) we get:

$$\tilde{f}(\zeta) = (2\pi i \zeta)^{-\beta} \int_{\mathbb{R}^n} e^{-2\pi i x \zeta} (\mathbf{D}^\beta \varphi)(x)\, dx.$$

Then

$$(2\pi i \zeta)^\beta \tilde{f}(\zeta) = \int_{\mathbb{R}^n} e^{-2\pi i x \zeta} (\mathbf{D}^\beta \varphi)(x)\, dx$$

and

$$\left|(2\pi\zeta)^\beta \tilde{f}(\zeta)\right| \leq \left(\sup_{|x|\leq R} \left|e^{-2\pi i x \zeta}\right|\right) \int_{|x|\leq R} \left|(\mathbf{D}^\beta \varphi)(x)\right| dx \, ,$$

now using Lemma E.1 on the left hand side of the inequality, we get for every $k \in \mathbb{N}$ a positive b_k such that

$$b_k \left(1 + |\zeta|^2\right)^k \left|\tilde{f}(\zeta)\right| \leq e^{2\pi R|\,\mathrm{Im}(\zeta)|} \left\|\mathbf{D}^\beta \varphi\right\|_1 \, .$$

So finally

$$\left(1 + |\zeta|^2\right)^k \left|\tilde{f}(\zeta)\right| \leq c_k e^{2\pi R|\,\mathrm{Im}(\zeta)|} \, .$$

(ii)$\Rightarrow$(i) : Suppose f is an entire function that satisfies (ii), and define

$$\varphi(x) = \int_{\mathbb{R}^n} e^{2\pi i x \xi} f(\xi) \, d\xi \qquad (x \in \mathbb{R}^n) \, .$$

Note first that $(1 + |\xi|^2)^{-N} f(\xi)$ is in $L^1(\mathbb{R}^n)$ for every $N \in \mathbb{N}$. Hence $\varphi \in C^\infty(\mathbb{R}^n)$ and f is the Fourier image of φ.
Now we need to prove that $supp\, \varphi \subset \{\, x : |x| \leq R \,\}$.
Let $\eta \in \mathbb{R}^n$ and let η_1 be the first coordinate of η.
Using the rectangle of vertices $-a$, a, $a + i\eta$, $-a + i\eta$, and following the argument in Example E.32, we have:

$$\int_{\mathbb{R}} e^{2\pi i(\xi_1 + i\eta_1)x_1} f(\xi_1 + i\eta_1, \xi_2, \ldots, \xi_n) \, d\eta_1 = \int_{\mathbb{R}} e^{2\pi i \xi_1 x_1} f(\xi_1, \xi_2, \cdots, \xi_n) \, d\eta_1 \, .$$

The same can be done for the other coordinates and then:

$$\varphi(x) = \int_{\mathbb{R}^n} e^{2\pi i(\xi + i\eta)x} f(\xi + i\eta) \, d\xi.$$

Using $|f(\xi)| \leq c_n (1 + |\xi|^2)^{-n} e^{2\pi R|\eta|}$, we get

$$|\varphi(x)| \leq c_n e^{2\pi(R|\eta| - x\eta)} \int_{\mathbb{R}^n} \frac{d\xi}{(1 + |\xi|^2)^n}.$$

Since the last integral converges, take $\eta = \alpha x$ with $\alpha > 0$ giving

$$|\varphi(x)| \leq C_n e^{2\pi\alpha|x|(R - |x|)}.$$

Thus, $\varphi(x) = 0$ if $|x| > R$ and $\alpha \to \infty$. So the support of φ is a subset of $\{\, x : |x| \leq R \,\}$. $\qquad\square$

The above theorem may be generalized to distributions with compact support. Let us start with a lemma that will be needed in the proof of the theorem. This is an abstract "differentiate under the integral sign" result.

Lemma E.50. *Let $D \subset \mathbb{C}^n$ be open, E a locally convex space, $\phi \colon D \longrightarrow E$ and $f \in E'$. Define $F \colon D \longrightarrow \mathbb{C}$ by $F(z) = \langle f, \phi(z)\rangle$.*

(i) If ϕ is continuous, then F is continuous.

(ii) If ϕ is differentiable at $z_0 \in D$, then F is differentiable at z_0 and
$$\frac{\partial}{\partial x_j} F = \left\langle f, \frac{\partial \phi}{\partial x_j} \right\rangle .$$

(iii) If ϕ is C^∞, then F is C^∞ with $\mathbf{D}^\alpha F = \langle f, \mathbf{D}^\alpha \phi \rangle$.

(iv) If ϕ is analytic, then F is analytic. (Here, differentiability means
$$\lim_{h \to 0} \frac{\phi(z + he_j) - \phi(z)}{h} = \frac{\partial \phi}{\partial x_j} \text{ with convergence in } E.)$$

Proof. (i) and (ii) follow from the continuity of f; (iii) follow from (ii). (iv) follow from the Cauchy–Riemann equations. $\square$

Example E.51. Let $e^{-2\pi i(\cdot)z}$ be the function $x \longrightarrow e^{-2\pi i x \cdot z}$ defined on $\mathbb{R}^n$. Then $z \longrightarrow e^{-2\pi i(\cdot)z}$ from $\mathbb{C}^m$ to $C^\infty(\mathbb{R}^n)$ is analytic. For each k the function $\phi_k \colon z \longrightarrow \displaystyle\sum_{j=0}^{k} e^{-2\pi i(\cdot)z}$ from $\mathbb{C}^n$ to $C^\infty(\mathbb{R}^n)$ is analytic and ϕ_k converges to the function $z \longrightarrow e^{-2\pi i(\cdot)z}$ in $C^\infty(\mathbb{R}^n)$.

Theorem E.52 (Paley–Wiener–Schwartz). *Let R be a positive number. The following are equivalent for a function f defined on $\mathbb{R}^n$:*

(i) f is the image of the Fourier transform of a distribution $T \in \mathscr{D}'(\mathbb{R}^n)$ with $supp\, T \subset \{ x : |x| \leq R \}$.

(ii) f can be extended to a function $\tilde{f}$ analytic on $\mathbb{C}^n$ with the following property: there exists an $m \in \mathbb{N}$ and $c > 0$ such that
$$\left| \tilde{f}(\zeta) \right| \leq c(1 + |\zeta|^2)^{\frac{m}{2}} e^{2\pi R |\operatorname{Im}(\zeta)|} \quad \text{for } \zeta \in \mathbb{C}^n .$$

Proof. (i)$\Rightarrow$(ii):
Let $f = \mathscr{F}(T)$ with $T \in \mathscr{D}'(\mathbb{R}^n)$ and $supp\, T \subset \{ x : |x| \leq R \}$. By Proposition E.48, $f(\xi) = \left\langle T_x, e^{-2\pi i x \xi} \right\rangle$ for $\xi \in \mathbb{R}^n$. Let
$$\tilde{f}(z) = \left\langle T_x, e^{-2\pi i x z} \right\rangle$$
for $z \in \mathbb{C}^n$. Using Theorem E.18 we see that $T \colon C^\infty(\mathbb{R}^n) \longrightarrow \mathbb{C}$ is continuous and then $\tilde{f}(z)$ makes sense. Our aim is to show that the function $\tilde{f}$ is analytic on $\mathbb{C}^n$. The function $z \to e^{2\pi i x z}$ is continuous from $\mathbb{C}^n$ to $C^\infty(\mathbb{R}^n)$

(Example E.51), and T is also continuous so $\tilde{f}(z)$ is the composition of two continuous function so it is continuous. Let's calculate the derivative on the first variable z_1. Let $b = (b_1, 0, \ldots, 0)$ we have

$$\frac{\tilde{f}(z+b) - \tilde{f}(z)}{b_1} = \frac{1}{b_1} \left\langle T, e^{2\pi i x(z+b)} - e^{2\pi i x z} \right\rangle.$$

Now

$$\frac{e^{2\pi i x(z+b)} - e^{2\pi i x z}}{b_1} \longrightarrow \frac{\partial e^{2\pi i x z}}{\partial z_1}$$

in $\mathcal{C}^\infty(\mathbb{R}^n)$ as $b_1 \longrightarrow 0$. Since T is continuous the second part of the equality tends to $\left\langle T, \frac{\partial}{\partial z_1} e^{2i\pi x z} \right\rangle$ and is equal to $\frac{\partial}{\partial z_1} \tilde{f}(z)$, the continuity of $\frac{\partial}{\partial z_1} \tilde{f}$ is clear and by induction we have that $\tilde{f}$ is analytic in $\mathbb{C}^n$.

Since T is continuous on $\mathcal{C}^\infty(\mathbb{R}^n)$ and has support in $\{\, x : |x| \le R \,\}$, there exist k and $c \ge 0$ that

$$|\langle T, \varphi \rangle| \le c \sup\{\, |\mathbf{D}^\alpha \varphi(x)| : |\alpha| \le k, |x| \le R \,\}$$

for $\varphi \in \mathcal{C}^\infty(\mathbb{R}^n)$ (Theorem E.18). Set $\varphi(x) = e^{-2\pi i x z}$. Then

$$|\tilde{f}(z)| = |\langle T, \varphi \rangle| \le c e^{2\pi R |\mathrm{Im}(z)|} \sup\{\, |(2\pi z)^\alpha| : |\alpha| \le k \,\}.$$

Finally applying Lemma E.1,

$$|\tilde{f}(z)| \le c e^{2\pi R |\mathrm{Im}(z)|} (1 + |z|^2)^{2k}.$$

(ii)$\Rightarrow$(i):
Let $\tilde{f}$ be a function such that

$$|\tilde{f}(z)| \le c(1 + |z|^2)^{\frac{m}{2}} e^{2\pi R |\mathrm{Im}(z)|}.$$

Take $x \in \mathbb{R}^n$ and define $f(x) = \tilde{f}(x)$ then

$$|f(x)| \le c \left(1 + |x|^2\right)^{\frac{m}{2}}.$$

Then using Example E.30, $f \in \mathcal{S}'(\mathbb{R}^n)$ and is the Fourier transform of some tempered distribution T. Let (ψ_j) be a sequence in $\mathscr{D}(\mathbb{R}^n)$ such that $\psi_j(x) \ge 0$ for all $x \in \mathbb{R}^n$, $\int_{\mathbb{R}} \psi_j(x)\, dx = 1$, *supp* $\psi_j \subset \{\, x : |x| \le \varepsilon_j \,\}$ where $\varepsilon_j \to 0$ as $j \to \infty$.

This sequence exists: take ψ the function defined by equation (E.3) and define $\psi_j(x) = \frac{1}{\varepsilon_j} \psi\left(\frac{x}{\varepsilon_j}\right)$ where $\varepsilon_j > 0$, $\varepsilon_j \to 0$. ψ_j will have the required properties.

Take $T_j = \psi_j * T$; then $\widehat{T_j} = \widehat{\psi_j} \widehat{T} = \widehat{\psi_j} f$. By the Paley–Wiener Theorem

E.49 ((i)$\Rightarrow$(ii)), each $\widehat{\psi_j}$ can be extended to an analytic function on $\mathbb{C}^n$ that we will again denote by $\widehat{\psi_j}$ which satisfies the following property: for all $k \in \mathbb{N}$ there exists $c_{k,j}$ such that:

$$|\widehat{\psi_j}(\zeta)| \leq c_{k,j} e^{2\pi\varepsilon_j |\mathrm{Im}(\zeta)|} \left(1 + |\zeta|^2\right)^{-\frac{k}{2}}.$$

Then for $h = k + m$ and $d_{h,j} = c \cdot c_{k,j}$ we have

$$|\widehat{T_j}(z)| = |\widehat{\psi_j}(z)f(z)| \leq c_{k,j} e^{2\pi\varepsilon_j |\mathrm{Im}(z)|} (1 + |z|^2)^{-\frac{k}{2} - \frac{m}{2}} c(1 + |z|^2)^{\frac{m}{2}} e^{2\pi R |\mathrm{Im}(z)|}$$

$$= d_{h,j} e^{2\pi(\varepsilon_j + R)|\mathrm{Im}(z)|} (1 + |z|^2)^{-\frac{k}{2}}.$$

Now using again Paley–Wiener Theorem E.49 ((ii)$\Rightarrow$(i)), T_j has compact support contained in $\{ x : |x| < R + \varepsilon_j \}$, when $j \to \infty$, $T_j \to T$ in $\mathscr{D}'(\mathbb{R}^n)$ (weak topology). Let $\varphi \in \mathscr{D}(\mathbb{R}^n)$ with $supp \, \varphi \subset \{ x : |x| > R \}$, since $supp \, \varphi$ is compact for j sufficiently big we have:

$$supp \, \varphi \subset \{ x : |x| > R + \varepsilon_j \}$$

and then $\langle T_j, \varphi \rangle = 0$ and since $T_j \to T$ we have $\langle T, \varphi \rangle = 0$ which proves that $supp \, T \subset \{ x : |x| \leq R \}$. $\square$

Corollary E.53. *The Fourier image of a nonzero distribution with compact support is never with compact support.*

Proof.   Let $T \in \mathscr{D}'(\mathbb{R}^n)$ with compact support and $f = \widehat{T}$. Then f can be extended to an analytic function $\tilde{f}$ on $\mathbb{C}^n$. So f is analytic on $\mathbb{R}^n$. But an analytic function in $\mathbb{R}^n$, that is zero in an open and non empty subset of $\mathbb{R}^n$, is identically equal to zero. So f cannot have compact support unless it is identically zero. $\square$

Bibliography

[An] R.F.V. Anderson, The Weyl functional calculus, J. Functional Anal. 4 (1969), 240-267.

[Ap] T. Apostol, *Mathematical Analysis*, Addison-Wesley, Reading, 1975.

[AZ] J. Alvarez, J. Zilber, *Cálculos Funcionales en Álgebras de Banach*, Fascículo # 31, Universidad de Buenos Aires, 1983.

[BN] G. Bachman and L. Narici, *Functional Analysis*, Academic Press, N.Y., 1966.

[Be1] S. Berberian, *Measure and Integration*, MacMillan, N.Y, 1950.

[Be2] S. Berberian, *Notes on Spectral Theory*, Van Nostrand, Princeton, 1966.

[Be] L. Bers, Introduction to Several Complex Variables, N Y University, 1964.

[Beu] A. Beurling, Sur les integrales de Fourier absolument convergentes et leur application à une transformation fonctionnelle, Congrés des Math. Scandinaves, Helsingfors, 1938, 345-366.

[Da] E. B. Davies, *Spectral Theory and Differential Operators*, Cambridge University Press, Cambridge, 1995.

[DM] L. Debnath and P. Mikusinski, *Introduction to Hilbert Space and Applications*, Academic Press, San Diego, 1990.

[DeS] J. DePree and C. Swartz, *Introduction to Real Analysis*, Wiley, N.Y., 1987.

[DU] J. Diestel and J. Uhl, *Vector Measures*, Amer. Math. Soc., Providence, 1977.

[Du1] N. Dunford, Spectral Theory I, Convergence to Projections, Trans. Amer. Math. Soc., 54 (1943), 185-217.

[Du2] N. Dunford, Spectral Theory, Bull. Amer. Math. Soc., 49(1943), 637-651.

[DS] N. Dunford and J. Schwartz. *Linear Operators I*, Interscience, N.Y., 1958.

[Ey] M. Eydenberg. The Weyl Correspondence as a Functional Calculus for Non-Commuting operators. Rocky Mountain J. Math. Volume 39, Number 5 (2009), 1467-1496.

[FL] W. Filter and I. Labuda, Essays on the Orlicz-Pettis Theorem I, Real Anal. Exch. 16 (1990/1991), 393-403.

[Gr] Grabiner, S., A short proof of Runge's theorem, Amer. Math. Monthly 83, 1976, 807-808

[Ha] P. Halmos, *Measure Theory*, Van Nostrand, N.Y., 1950.

[Ho] J. Horvath. Topological Vector Spaces and Distributions. Addison-Wesley, 1966.

[HP] E. Hille and R. Phillips, *Functional Analysis and Semi-Groups*, Amer. Math. Soc., Providence, 1957.

[HS] E. Hewitt and K. Stromberg, Real and Abstract Analysis, Springer-Verlag, Berlin, 1965.

[Ka] N. J. Kalton, The Orlicz-Pettis Theorem, *Contemporary Math.* 2, Amer. Math. Soc., Providence, 1980.

[KK] I. Kluvanek and G. Knowles, *Vector Measures and Control Systems*, North Holland, Amsterdam, 1975.

[Kl] I. Kluvánek, Miery v Kartezsych Sucinoch, *Cas. Pest. Mat.* 92 (1967), 282-286.

[Or] W. Orlicz, Beitrage zur Theorie der Orthonalent wichlungen II, *Studia Math.* 1 (1929), 241-255.

[Ne] E. Nelson, Operants: A functional calculus for noncommuting operators in functional analysis and related fields, Proceedings University of Chicago, May 1968, Springer Verlag 1970 p.172-187

[Pe] B. J. Pettis, Integration in Vector Spaces, *Trans. Amer. Math. Soc.* 44 (1938), 277-304.

[Ri] F. Riesz, Les systèmes d'équation linéares à une infinité d'inconnues, Paris, 1913.

[RN] F. Riesz and B. Nagy, *Functional Analysis*, Unger, N.Y., 1955.

[Roy] H. L. Royden, *Real Analysis, Third Edition*, Prentice-Hall, Eaglewood Cliffs, 1988.

[Ru1] W. Rudin, *Functional Analysis*, McGraw Hill, New York 1973.

[Ru2] W. Rudin, *Real and Complex Analysis*, 3rd. ed., McGraw Hill, New York 1974.

[RT] L. A. Rubel and B. A. Taylor, Functional analysis proofs of some theorems in function theory, Amer. Math. Monthly 76 (1969), 483–489

[Ste] L. Steen, Highlights in the History of Spectral Theory, *Amer. Math. Monthly* 80 (1973), 359-381.

[Sw1] C. Swartz, Measure, *Integration and Function Spaces*, World Sci. Publ., Singapore, 1994.

[Sw2] C. Swartz, *An Introduction to Functional Analysis*, Dekker, N.Y., 1992.

[Sw3] C. Swartz, *Multiplier Convergent Series*, World. Sci. Publ., Singapore, 2009.

[SZ] Saks S., A. Zygmund, *Analytic functions*, 3rd. ed. Elsevier, Amsterdam, London, New York, 1971.

[Ta1] A. Taylor, Spectral theory of closed distributive operators, Acta Math., 84(1950), 189-224.

[Ta2] A. Taylor, Historical Notes on Analyticity in Functional Analysis, Problems in Analysis, Princeton, 1970, 325-343.

[Tay] M. E. Taylor, Functions of several self-adjoint operators, Proc. Amer. Math. Soc. 19 (1968), 91-98.

[TL] A. Taylor and D. Lay, *Introduction to Functional Analysis*, Wiley, NY, 1980.

[Th] G. E. F. Thomas, L'integration par rapport a une mesure de Radon vectorielle, *Ann. Inst. Fourier* 20 (1970), 55-191.

[Wh] R. Whitley, The Spectral Theorem for Normal Operators, *Amer. Math. Monthly* 75 (1968), 856-861.

[Wie] N. Wiener, The Fourier Integral and Certain of its Applications, Cambridge, 1933.

[Yo] K. Yosida, Functional Analysis, Springer-Verlag, Berlin, 1965.

Index